William Paul Gerhard

Recent Practice in the Sanitary Drainage of Buildings

Second Edition

William Paul Gerhard

Recent Practice in the Sanitary Drainage of Buildings
Second Edition

ISBN/EAN: 9783744678902

Printed in Europe, USA, Canada, Australia, Japan

Cover: Foto ©berggeist007 / pixelio.de

More available books at **www.hansebooks.com**

RECENT PRACTICE

IN THE

SANITARY DRAINAGE

OF BUILDINGS,

WITH

MEMORANDA ON THE COST OF PLUMBING WORK.

BY WM. PAUL GERHARD, C. E.,

Consulting Engineer for Sanitary Works.

(New York City.)

SECOND EDITION, REVISED AND ENLARGED.

NEW YORK:
D. VAN NOSTRAND CO.,
No. 23 MURRAY AND 27 WARREN STREETS.

1890.

Copyright, 1890,
D. VAN NOSTRAND COMPANY.

PREFACE.

A great philosopher once said that "the fault of most books is their being too long." The author's aim has been to avoid this fault, by stating in plain language, as briefly as possible, what constitute the leading requirements of sanitary drainage as applied to buildings. The first and third parts (the second, third and sixth in the new edition) of this little book deal with the general principles of house drainage, while the second part (the fourth in the new edition) discusses in detail the requirements as to material, workmanship and arrangement of sanitary plumbing. These, it is believed, will be found particularly useful by architects and engineers in preparing complete plumbing specifications for all classes of buildings. The fourth part, (the fifth part in the new edition) giving memoranda on the cost of plumbing work, will, it is hoped, be found by many to be a welcome addition.

It is suggested that those who desire to obtain a complete knowledge of this important branch of interior house construction should read, in connection herewith, the author's former works, which this volume is merely intended to supplement.

THE AUTHOR.

39 UNION SQUARE, WEST,

NEW YORK CITY, May 1, 1887.

PREFACE TO SECOND EDITION.

The new edition of this book has been thoroughly revised and enlarged by the addition of two essays, one on "*Architecture and Sanitation*," the other on "*The Drainage of a House.*"

WM. PAUL GERHARD.

39 UNION SQUARE, WEST,

NEW YORK CITY, Sept. 1st, 1890.

CONTENTS.

Page.
I. ARCHITECTURE AND SANITATION............ 9

II. RECENT PROGRESS IN HOUSE DRAINAGE AND PLUMBING..................... 21

III. THE DRAINAGE OF A HOUSE................ 59

IV. MAXIMS OF PLUMBING AND HOUSE DRAINAGE................................ 89

    A. Rules regarding the Placing and Location of Plumbing Work in Dwellings............................ 94

    B. Rules regarding the Proper Construction of the Work..................... 98

        1. General Conditions............... 99
        2. Materials 102
        3. Workmanship 115
        4. General Arrangement of Plumbing Work.....................127

Page.
5. Tests of the Work during Construction and after Completion....143

C. *Rules regarding the Proper Care and Management of Plumbing Apparatus*.145

V. MEMORANDA ON THE COST OF PLUMBING WORK 153

VI. SUGGESTIONS FOR A SANITARY CODE.......163

A. Rules as to Healthful Building Construction........................163

B. Rules as to Connection between House Drains and Street Sewers...165

C. Plumbing Regulations............ ...169

I.
ARCHITECTURE
AND
SANITATION.

ARCHITECTURE AND SANITATION.

Every architectural structure, to be worthy of the name, should fulfill the following cardinal requirements: It should be truthful in expression and style, beautiful in design and form, and healthful, durable, convenient and practical in construction and arrangement.

A building should be truthful in character, not merely as regards the fitness of all interior arrangements for the purposes to which it is to be devoted, but its exterior should at once distinctly characterize its object and purpose. True art, further, demands truthfulness in construction and in the relation between the interior and the exterior of the building, as well as in the choice and employment of such building materials as are best adapted to the various parts that make up the structure.

Next in importance to truthfulness is the requirement of harmony and order, not always the mere symmetry of arrangement, but a general beauty of design as expressed by the peculiar characteristics of the building materials selected, by the architectural order, by the harmony between the details, between the parts and the whole ; in other words, by the proportions chosen between length, width and height of the structure, between the solid walls and the openings for doors and windows, and other details.

The foregoing are what we may term *ideal* requirements in distinction to the mere *practical* ones. According to the character and purpose of a structure, either the artistical or the practical requirements will predominate, but an architect, to be successful, must be familiar with both the æsthetic and the constructive principles which should guide in the conception, planning and designing of buildings ; he must be not only an artist, but a constructor as well.

To be skilful as a constructor he must have a thorough knowledge of all building materials, of the strength and stability of structures, of the different methods and systems of construction, and of the practical work performed by the various building trades. He must have a large and varied experience derived from a personal superintendence of buildings of different character, design and construction. Last, though not least, he should be familiar with the means available to protect a structure against the hurtful influences of the elements, and with the best methods obtainable for rendering a building healthful, for no building which is to be the abode of human beings, or even animals, however practical and convenient in arrangement, and however beautiful in design and ornamentation it may be, can be considered *perfect,* unless it is entirely free from any influence which may, directly or indirectly, injuriously affect the health of its inmates. It matters little whether or not the origin or spread of certain diseases can be *proven* to be due to unsanitary conditions or surroundings. Of the exact mode in which filth acts upon the human system we know as yet very little. What we do know, however, with a tolerable degree of certainty, is that much sickness, suffering and premature death may be prevented by the removal of all influences injurious to health.

To find the above outlined requirements of artistic genius, scientific knowledge and practical skill and ability combined in the person of one man is rare indeed, for in architecture, as elsewhere, novel problems constantly arise, increased civilization and luxurious modes of living continue to create new wants, and the whole subject is growing to such a vast extent that it can hardly be thoroughly mastered by a single person, even in a lifetime. It would, therefore, seem possible, and, indeed, it has been done in many cases, to effect a division of labor.

Comparatively few in number are those structures where the artistic requirements, the ornamentation and decoration predominate. Much more numerous are buildings where the practical requirements of the design—sometimes, indeed, the necessity of the utmost economy in construction—demand the entire ignoring of all merely decorative features. It is not necessary to quote examples, for instances of either class of buildings will easily recur to the reader's mind. The majority, however, of all architectural structures, no matter what the object may be to which they are devoted, require the equally strict observation of both ideal and practical requirements, and we venture to say that the best results will often be attained by a combination of the highest talents both in design and in construction.

Whether a building is intended as a transient or permanent domicile for mankind, its perfect salubriousness is certainly equal in importance to the requirement of suitable design and arrangement, and of durable and sound construction. The protection of a structure against the influence of the climate and seasons, of heat and cold, against wind and rain, against fire and lightning, against dampness or dangerous exhalations from the soil, rightfully demands a profound regard and the most careful consideration, and to this should be added the necessity of providing suitable arrangements to supply buildings with light and air, water and heat, and to remove from them the many waste products incident to human life. All the latter exigencies are eminently of a utilitarian character, and have not always, hitherto, received the attention due them. This may be partly explained by the fact that researches and investigations relating to practical sanitation have only recently been undertaken. It is pre-eminently due, however, to the almost universal indifference toward the subject, and to the heretofore existing demand of the public for either handsome buildings or very cheap houses.

We have, in the past few years, witnessed a remarkable change in this respect, and the once indifferent public, having become aware of the danger to health arising from, and the many diseases caused by, living in unhealthy houses, begins to appreciate the importance of the question, and architects, in their turn are obliged to pay more attention to the subject.

Owing, probably, to the purely practical aspects of the question, the latter has become the special province of the sanitary

engineer, whose energy is devoted to the
contriving of the best means for preventing
a pollution of the soil, the air, or the water
about habitations, and who, wherever
habitations are closely grouped together,
as in villages, towns and large cities, devises and carries out schemes of drainage,
water-supply and sewerage. It is but natural that, having provided means for the
establishment of healthful conditions about
buildings, he should go a step further and
devote close study and attention to the
securing of healthful conditions *inside* of
habitations. These, indeed, are of even
greater moment to the individual, for of
the two evils, a healthy home amidst unhealthy or an unsanitary home in healthful surroundings, the latter is obviously
the one capable of working more direct
harm.

We thus find that the essential, though
to many uninteresting, details of water-supply, house drainage, plumbing, heating,
ventilation and lighting, and healthy
foundations, have of late become the particular sphere of the sanitary engineer.

That this should be so is not, however surprising, if we remember what was previously asserted about the magnitude of the subject with which architects have to deal. We do not wish to be understood as meaning that an architect should not devote attention to questions relating to healthful house-construction. But, we hold the opinion that, at least in the case of important structures, the architect should call in the aid of the sanitary engineer, particularly so because the problems of external and internal water-supply and sewerage are often so correlated and closely allied to each other that they cannot well be treated separately. In the construction of large buildings, and particularly whereever healthfulness is a prime consideration, it has become quite a common practice with architects in Europe to associate with themselves a sanitary engineer. The same practice is spreading in this country, and the writer has had the honor of having been associated in the past years with many of the most prominent architectual firms in this country, on large as well as small structures.

By such a division of labor, or such a combination of genius, talent and skill, it would seem to be easy to secure to a structure at once the best artistic, constructive and sanitary features.

Since the above paragraphs were penned the author's views have been corroborated by many architects of the highest standing. Quite recently Mr. Burnham, of the eminent firm of Burnham & Root, Architects, in Chicago, made some remarks upon the practice of architecture before the Chicago Architectural Sketch Club, which so fully endorse the views held by me that I make no apology for quoting portions from the June, 1890, issue of the *Chicago Inland Architect:*

"Buildings are now supplied with complex heating, plumbing, sewerage, ventilation, elevators, pumps, tanks, engines, dynamos and electric lighting. Many have costly plants of all these kinds, none are without a part. The owners have a right to expect the very best of everything in every part, and perfect mechanical planning and application in each case. Therefore, very expert mechanical and sanitary engineering are required to handle this part of the work.

"If many important works be on hand in an office at once, the mere handling of the business they produce requires much experience, and is enough to occupy the time of an able man.

"Passing over many other strong points that might be instanced, I have shown enough to prove tha one man cannot himself attend to all the work of a modern city practice in America, and as we agree that attention to details is necessary to success, it is plain that some method must be found to attain it, other than through a wider practice by a single person in the old way.

"It is manifest that the successful practitioner of the future will be a specialist, or at the head of an organization of specialists.   *   *   *   *

"To design and construct a great office building, some stores and flats, some dwellings, some factories, a church, etc., in a manner to elicit praise from our critical building public, is beyond the powers of one man's brain and energy, if he attend to the details of all. If, therefore, a man or a firm is ambitious to carry on a great general practice, there must be in the organization a very great designer, an exceptionally strong chief engineer, a sanitary engineer, a mechanical engineer, and a business man Each of these will have his hands full if he is faithful, and only with such an organization, I say again, can a large general practice live and keep going."

II.
RECENT PROGRESS
IN
HOUSE DRAINAGE
AND
PLUMBING.

RECENT PROGRESS
IN
HOUSE DRAINAGE AND PLUMBING.*

Among the many practical and utilitarian details of interior construction tending to increase the comfort and convenient arrangement of houses, none occupy a more important position than those relating to the fixtures, traps and pipes which introduce and distribute into our buildings a supply of pure water for household use, and afterward remove from them the liquid and semi-fluid foul wastes, designated by the general term house sewage.

Our modern homes present, in the vast number of pipes of all sizes, kinds and character which traverse them in all directions, not only across the basement or cellar, but also from cellar to garret, an

---

* This paper was originally prepared for, and appeared in, the Chicago *Inland Architect.*

appearance quite different from the houses built by our ancestors a century or more ago. A modern residence fairly abounds with pipes for the conveyance and distribution of illuminating gas into rooms and halls; with hot and cold water pipes; with pipes for heating purposes, carrying steam from boilers, or returning to them condensed water; with channels or conduits carrying fresh air into the cellar, to the heating apparatus, or directly into living or sleeping rooms; with flues and registers for the admission of pure, warm air, ventilating flues for the removal of vitiated air, and chimneys for carrying off smoke from the combustion of fuel; with speaking tubes, telephone wires, tubes enclosing wires for electric bells and for electric lighting; and finally with waste and vent, soil and drain pipes for the removal of household wastes through plumbing appliances, which are more or less scattered in all directions over the principal floors of the house.

These are what the tempting language of the advertisements of shrewd real estate agents or speculative builders comprise under the term "*modern improvements,*" and the minds of American householders have been accustomed to the latter to such an extent that even small houses rarely remain without some of the above named comfort-promoting arrangements; and that a much higher price is willingly paid for the purchase or lease of a building well provided with them. Of course, with lighting and heating apparatus of all sorts and kinds in a house, requiring intelligent care and frequent inspection, but generally left to be manipulated by ordinary servants, and with the still prevalent custom of having much of the work concealed, partly or wholly, in walls, partitions or between floor-beams, repairs are apt to be numerous, disagreeable and costly; and to avoid them, even in the early days of the introduction of such improvements, architects, builders and owners rarely failed to specify certain requirements, or neglected to make use of certain tests, to assure themselves that the pipes intended to convey water, steam, hot air, fire and smoke, and illuminating gas, were tightly jointed, so as to prevent the unwelcome or unwholesome escape of any of their contents into the house.

But with sewer pipes, soil and vent pipes, fixtures and traps, the usual practice has been different. The requirements in the majority of instances were less stringent, the danger arising from defective work was disregarded or overlooked through ignorance, indifference and folly, and the owner was usually less inclined to spend a sufficient amount of money for such work, except so far, perhaps, as the desire went to make a handsome display of richly decorated bowls, plated or gilt faucets, and elaborate ornamental woodwork, corresponding in style and elegance to the artistic decoration and interior finish of the rooms and richly adorned outside of the building. It was idle, in those days, to attempt to speak to the builders of a *sanitary* arrangement of plumbing work. Architects, consequently, paid less attention to the subject than it rightfully demanded.

But the world moves, and, although

progress in the art of properly draining houses seems to have been rather slow, yet in the past few years a decided change for the better has been noticeable. A stir has been made in the interest of healthy homes and healthy living largely by the dissemination of useful knowledge on the subject through newspapers, popular magazines, sanitary journals, pamphlet and health reports. The indifference of the public gradually vanished, sanitary laws began to be better understood, and the evil influence upon health of faulty plumbing work became apparent. A loud demand for healthful houses arose, sanitary surroundings were eagerly sought for, and the consequence was that architects, to meet the wishes of their clients, were obliged to give the subject some thought and attention. Builders, in their turn, were not slow to guard their pecuniary interest by, at least a superficial endeavor to offer what the public demanded, and hereafter *sanitary* plumbing and *sanitary* heating became the leading features of many advertisements of speculative house builders.

The greatest step forward, however, was made when in our large cities plumbing work became subject to board of health rules and regulations. At first there was, as might have been expected, considerable ill-feeling against compulsory measures, especially against the intrusion of the ever-vigilant sanitary inspector into the interior of private houses. The sacredness of personal rights, as embraced in the old saying "My house is my castle," was apparently violated. But a moment's consideration of the vast multitude of houses built as cheaply as possible, to sell or rent as profitably as possible, will convince any one how necessary such inspection must be. By way of digression, we would like to see a similar inspection enforced by law regarding the fireproof construction of buildings. Dwellings and apartment houses are frequently sold or rented as "*fireproof*," when in reality they are little better than tinder-boxes, falling a quick and easy prey to the flames of a once-started conflagration.

The beneficial effect of plumbing laws

extends principally to the vast number of householders and families in large cities, who are compelled to live in hired dwellings, apartments, or tenements, and who cannot protect themselves against preventable sickness due to unsanitary surroundings, in particular to defective drainage. The standard of plumbing work in such buildings, where all items of expense are apt to be reduced to a minimum, became appreciably improved and the condition of healthfulness increased, by giving its control to a board of health. Even if no other good results had followed the enforcement of rules and the official inspection of all work in new buildings, the single measure compelling the owner of each new house to file plans and specifications clearly showing and describing the system of plumbing to be introduced into the building, had a wholesome effect upon the manner in which such work, henceforth, was to be handled by architects as well as plumbers. Since no work could be commenced in any building before the plans and specifications filed had been approved by the board of health, the subject of drainage and plumbing received its share of proper attention at an early stage of the construction of the building.

The benefits derived from these measures have, it is true, by no means remained without some drawbacks, which, though not necessarily fatal to the results, still, to some extent, hinder as rapid a progress as might have been expected. To these I desire to make a brief allusion. So far as owners are concerned, we now often find too much reliance placed by them in the supervision exercised by boards of health. Even with a multitude of efficient inspectors in their employ, this supervision must, of necessity, be very general only, and with the still universal tendency of covering up and burying out of sight those "unsightly pipes and fixtures" much defective work *may* escape the eye of an Argus-eyed and faithful inspector. There is, moreover, in every plumbing job a chance for much botched work, not necessarily involving a violation of any plumbing regulation. On the other hand, since some points relating to house drainage are yet disputed, it cannot be expected that plumbing regulations will become perfect for some time to come. Even the most complete regulations which I have seen contain certain rules and requirements which appear to me to be of doubtful value. Considerations towards manufacturers of plumbing goods will generally prevent boards of health from prohibiting the use of appliances which all sanitarians have long ago condemned. Again, it is obvious that such general regulations must be quite inadequate and incomplete when applied to work in very large and extensive structures. They cannot include rules regarding all the details of the work, for, as is natural, board of health laws can only insist upon and enforce a certain minimum of improvements.

As regards architects, judging from a somewhat extensive personal experience with a number of plumbing specifications with which I came in contact during the past few years through the practice of my profession, much improvement is noticeable, together with a desire to keep well informed about the progress in the principles and the practice of house drainage. Yet the majority of their specifications continue to be written in a fashion-like manner, without sufficient attention to the details and to improved methods and appliances, and failing to lay proper stress upon the quality of the materials, upon the workmanship required, and upon proper and stringent tests regarding the pipe system. Nor is the inspection of such work a thorough one, and the testing of the pipe system by pressure, if specified, is rarely insisted upon by architects. In those cities where boards of health require the filing of plans and specifications, it became customary with architects to use blank specifications, printed and furnished by the health board, which contain as a guide the essential requirements, the blank spaces in them being simply filled out by architects, who would add a few additional requirements in writing. I do not know whether or not the use of such blank specifications is made compulsory, in which case no blame could fall upon architects for using them. What I do know is that such specifications, as applied to a large building, must necessarily be quite incomplete, insufficient and imperfect, and often appear unsystematic in the arrangement of the subject matter.

The object of a house drainage system has long been understood to be to remove from dwellings at once, and as thoroughly as possible, all liquid waste matters before they undergo decomposition and emit unwholesome gases of putrefaction. In its widest meaning the term "house drainage" is made to include the removal of surface water from roofs, areas and yards, as well as of subsoil water from the ground upon which a building is erected. Strictly speaking, we should distinguish between *drainage* and *sewerage* of a building, the former term referring to the removal of all clear water (roof water, surface water and subsoil water), the latter to that of fouled waters from the household (sewage proper). It is often of very great importance to keep the one separate from the other, as, for instance, in the case of isolated country houses, where sewage is conveyed by *tightly-jointed* sewer pipes to a flush tank or other tight receptacle, from which it is disposed of by irrigation or sub-surface irrigation upon or underneath lawns or grass land, while subsoil water is removed by *open-jointed* land drains to the nearest watercourse, brook or open ditch. A third system of pipes generally carries, in the case of rural dwellings, the rain falling upon the roof into a tight cistern or storage tank for clean water. With city houses it is sometimes, though not often, feasible to carry out the same separation, at least as far as the subsoil water is concerned, provided a special line of subsoil drains has been laid in the street in the same trench with the sewer. As a rule, in the case of narrow city lots, one outlet is common to the drain and the house sewer, but it is one of the most important duties of those who advise in such matters to consider the best means for safely disconnecting the open-jointed tiles from the house or street sewer, to prevent the back flow of sewage and—still more important — of sewer air from the house sewer into the tile drains, from where it would easily diffuse into the cellar and rise to the upper floors of a house.

The removal of subsoil water, or at least the permanent lowering of its water level to secure dryness of the house, is equal in importance, from a sanitary point of view, to a proper system of house sewerage and plumbing. It is best accomplished by lines of small-sized common, porous land drains, laid at least two feet below the cellar floor, with open joints, protected by a collar, wrapping paper or muslin, against chokage from dirt, all delivering into a main drain, sloping toward the outlet, which may be either a gutter in the road, an open ditch or watercourse, or else the sewer in the street. In the former case no further protection is needed at the outlet, except a strong, fixed grating, to prevent the entrance of mice or rats. Where subsoil water is discharged into sewers an efficient disconnection should be provided by a deep-seal water trap, kept constantly filled by some automatic device in connection with the water supply of the house.

In advising clients in regard to subsoil drainage, I generally call their attention to a few points regarding healthful house construction, not strictly belonging to, but intimately connected with, the ventilation and drainage of dwellings, and which, in most cases, remain unheeded by architects and builders. I refer to the perfect isolation of the house from the ground upon which it stands, and from the water and air contained in the pores of the soil. It is not enough to provide for the subsoil drainage under the house; its walls are frequently exposed to moisture, or even to water veins which are penetrated in digging trenches for foundations. Dampness of walls is a frequent occurrence in cheaply built houses as well as in the better class of houses where no attention is paid to such apparently insignificant details. Again, the damp vapor of ground air and frequently unhealthy exhalations from polluted soils would constantly rise into the cellar, especially if assisted in their upward passage by the so-called "suction" of house chimneys, unless provision is made for a thorough isolation of the cellar from the ground below by a tight cellar floor, which at the same time will prevent the rise of subsoil water. Asphaltum has proved a very valuable material for foundation walls, damp-proof courses, as well as for the cellar floor, and efficiently accomplishes this much to be desired complete isolation.

A safe drainage of the subsoil being arranged for, we must next provide for the speedy and complete removal of the house sewage, consisting of waste water from flushing urinals and water closets, together with human excrements and urine, of dirty water from personal ablutions in washbowls and all forms of bathing tubs, of chamber slops, of foul laundry water and water used for rinsing cooking vessels and cleaning dishes. The amount of sewage will be largely increased at times if the rain water falling upon the roof and upon paved areas and court-yards is also admitted into the house sewer. Whether or no this should be done will depend upon the system of sewerage existing in the place, but even where the street sewers are designed to carry and to ultimately receive more or less rainfall, the question arises whether it is better to have within the house a single system or a double set of pipes, one for sewage and another for rain water. This question cannot be decided in a general way. It becomes necessary to take into consideration the special conditions speaking for and against such separation, and thus each building becomes a problem in itself. This much may, however, be stated, that it is preferable to keep the vertical pipes leading the water from the roof separate from vertical soil or waste pipes.

In a brief article on the subject it is, of course, impossible to refer to the many details of plumbing work.* I must necessarily restrict myself to a statement of the *leading requirements* and *general principles* governing the planning and arrangement of such work. What, in the light of present definite knowledge of the subject, I consider as essential, and as applicable *without modification* to all classes of buildings, to the drainage and sewerage of the largest and most expensive mansion, the smallest city house or suburban cottage, of schools, hotels, hospitals, factories and tenement houses, may be summarized as follows :

I should use within a building metal pipes only, principally iron pipes, with the exception of the short branch waste and supply pipes, which may be of lead. I should commence with the iron drain at

---

* The reader may find the subject thoroughly discussed in the author's works : — House Drainage and Sanitary Plumbing, 2d Edition, 1884.

Hints on the Drainage and Sewerage of Dwellings, 2d Edition, 1884.

Guide to Sanitary House Inspection, 3d Edition, 1890.

Domestic Sanitary Appliances (in press).

See also the article " Maxims of Plumbing and House Drainage," in this book.

least five feet beyond the foundation walls, so as to make sure against breakage by settlement of walls, and to further guard against the latter serious calamity I should always advise to turn a relieving arch across the wall where the pipe passes out. In the case of country houses, with a water supply derived from a well in the vicinity of the house, I should advise carrying the house sewer of iron to a point well beyond the probable limits of the drainage area of the well. I should recommend, even in those houses where due regard to economy must be had, the use of extra heavy pipes of uniform thickness and tested under pressure before use, and before applying a protective coating of coal tar or a similar substance, so as to avoid any imperfections in the pipe which may be covered up by the enameling or tarring process. I should also insist, at a later stage of the work, upon a proper test of the pipe joints in order to make sure that the whole system is air-tight as well as water-tight beyond any doubt. I should advise the use of a diameter of only four inches for the main pipe of a single house of ordinary size, and should restrict the size of the main drain of larger buildings to five and six inches, preferring to arrange two or more systems of six-inch main sewers for the largest institutions, in place of one eight or ten-inch pipe.

I should, wherever possible, banish all plumbing fixtures from the cellar floor, in order to carry the house drain in plain sight either along one of the cellar walls, or else suspended from the basement floor beams.

I should give to the pipes in the cellar all the fall possible, in order to secure a good cleansing velocity of the flow in the main pipe, and wherever the needed fall could not be obtained I should advocate the use of flushing tanks of some kind at the head of the drains. I should suggest strongly to support the main drain and its branches by brick piers, placed at suitable intervals, and especially at the junction of all upright pipes. I should take care to have all junctions made with Y branches instead of T branches, and all changes from the direct line made with curves of an easy sweep. I should recommend the use of cleaning hand-holes, at intervals, along the main line, at junctions, bends, and near traps, but I should also strongly counsel the thorough and tight closing of all such inspection openings.

I should carry all upright soil pipes, and all lines of waste pipes, in the straightest practical course, and with as few elbows as possible, up to and through the roof, and should advise making this extension in no case less than four, and preferably six inches diameter, to provide a free outlet above the roof. This outlet I should carry well above the roof line, and should keep it away as far as possible from any chimney flues, ventilating shafts, dormer windows, etc. I should firmly insist upon a copious and constant circulation of fresh air through all drain, soil, waste and vent pipes, and with this end in view, should provide a suitably large inlet for air, at the lowest point of the system, and extend all pipes at least full size above the roof, doing away entirely with any obstructions in the shape of ventilators, cowls, caps, or, worst of all, return bends, covering the mouths of pipes. If there is ground to fear an accidental or malicious obstruction of the pipes, I should urge the use of only a wire netting, or a common leader guard inserted into the pipe mouth, or, what is better, I would extend the pipes sufficiently high to keep their open mouth out of reach of mischievous persons.

I should recommend locating all fixtures as much as possible in vertical groups in order to get a straight, simple, and direct arrangement of soil and waste pipes, and to reduce the length of branch waste pipes, thereby securing a more thorough and direct discharge of fixtures. I should not use a soil pipe larger than four and five inches inside diameter, even for the greatest possible number of fixtures, and I should limit the size of upright waste pipes for sinks, basins or baths to two and three inches.

Regarding plumbing appliances, my advice would always, even in the case of the most costly residences, consist, for obvious reasons, in reducing their number, and consequently the amount of plumbing work, as much as practicable, and to avoid placing fixtures in spare rooms, where they would not be constantly used. I should further recommend to locate all fixtures in well-lighted and well-ventilated rooms, thereby insuring a proper use, and a better care of the appliances. I should abolish all plumbing from sleeping rooms, confining the same to the bath room, the kitchen, the laundry, the pantry, and to well-lighted closets.

Although the subject of warming and ventilation cannot be here considered, yet I will mention that I should insist upon a proper and constant change of air in the bath rooms and water closets. This involves the introduction and thorough diffusion of an ample supply of pure air from outside, moderately warmed (in our climate during at least seven months of the year), not only to increase the comfort of the bath room, but also to prevent the freezing of supply pipes, or of the standing water in traps. It also requires the removal of the foul air, which can be attained in a simple, yet efficient manner by arranging a gas-burner in an outlet flue of ample size. These are matters which begin to be better understood in the construction of houses, but I desire to call attention to a defect which I have frequently noticed in otherwise well-ventilated houses, namely, that where a strong suction exists from the outlet-flues or chimney-places provided in rooms, halls, or staircases, the supply of air, and sometimes noisome odors, are frequently drawn from a bath room or a slop closet. Hence it should be borne in mind, in arranging a general system of house ventilation, that the ventilation of apartments containing water closets, urinals, slop hoppers, or other fixtures, requires special attention, and that to be effective and reliable there should, preferably, be a constant movement of air from the other parts of a house toward and into the bath room; in other words, it is of prime importance to arrange a well-drawing outlet flure in a bath room water closet apartment, which would tend to create a slight vacuum in said room. Sufficient air being thus constantly removed, fresh air will easily come in to take its place, provided it is admitted in ample quantity into the other parts of a house. Where a building is ventilated by *plenum ventilation*, it is better not to include bath-rooms or water closet apartments in such a system.

It should be the aim to have the whole plumbing work arranged as simply as possible. Supply pipes must always be so located that they will not freeze in cold weather, and it is preferable to keep them away from the outside walls, unless special protection is given them. Householders, having lived during a winter in an exposed country house, are always ready to appreciate measures tending to the protection of water pipes against frost.

Whenever I am left untrammeled by prejudice I always arrange all plumbing work in an open manner, leaving all appliances, traps, supply and waste pipes *fully exposed to view*. The advantages gained hereby are two-fold. In the first place, I secure a better and more thorough workmanship of those parts of the work, which, being usually tightly boxed up, are very apt to be less carefully finished, and this is true not only of the plumber's work, but also of that of the carpenter and plasterer.

From a sanitary point of view, and likewise for other reasons, it is quite important to have all those unsightly holes where pipes pass through floors and ceilings tightly and permanently closed, to prevent diffusion of air from one story to another. I recently examined a bath room in an apartment house, where at each cold spell such a violent draft was rising through a pipe channel leading from the cold basement to the upper floors, along the soil and supply pipes, as to completely chill, and cause the freezing up of the water in the pipes, much to the annoyance of the house-owner, who could hardly get along without the plumber as soon as the thermometer would reach the freezing point, yet the plumbing in this building was done "in accordance with all the board of health rules" and had successfully passed inspection.

A second advantage obtained by leaving plumbing work fully exposed to view is that there is a better circulation of air around the fixtures, that the cleaning and scouring operations of servants are much facilitated, that all parts of the work are easily accessible and readily inspected, and that repairs are less frequent, and if they become necessary that there is little or no tearing up of wood work, floors, and base boards. An open arrangement also aids in enlightening the anxious minds of some householders concerning the "hidden mysteries of plumbing work." What is true of plumbing fixtures is, of course, equally applicable to the system of pipes in a dwelling. I strongly advise keeping all pipes outside of walls or partitions, locating them, where possible, in closets, or in inferior rooms. This enables one to inspect at any time any pipe joints or to readily reach any stop-cock or valve, should it be necessary to shut off the water from any pipe. I generally dispense with unsightly lead safes under fixtures, believing that with the open arrangement a leakage cannot remain unnoticed for a sufficient length of time to work serious harm, especially where walls (to the usual height of wainscoting) in kitchens and bathrooms are made water tight, and finished in tiles, plain or ornamental, or enameled brick, and where floors are finished with marble, tiles, slabs of slate, cement, or in simple *terrazzo* work. If required I arrange a drip pipe to remove any water from leakage, which pipe must always be kept entirely disconnected from any soil or waste pipe.

As to the fixtures proper, I should select for an inexpensive cottage, as well as for a luxuriously furnished city residence, those of a simple character, with a smooth, and non-absorbent surface. The exact material of the fixtures is often mainly a question of cost. For water closets, slop hoppers and urinals, which latter, however, I avoid in private houses, I should give preference to those with a small fouling surface, made in annealed glass or in earthenware. I recommend, of course, using water closets without any mechanism or moving parts liable to get out of order. I would, wherever I could, avoid the use of fixtures requiring a hidden overflow pipe. Bath tubs of all kinds, wash bowls, pantry sinks, water closets and urinals may now be had of such a form and construction as to do away entirely with concealed overflow channels, which are often the cause of annoying odors.*

I should locate fixtures as near as practicable along a soil or waste pipe, to avoid the always objectionable branch wastes under floors. I should endeavor to place the fixtures of different floors in groups arranged as nearly as may be vertically above each other, to reduce the number of pipe stacks. I should also aim to give to each fixture an independent and direct discharge into the vertical pipe. I should insist upon the separate, safe and secure trapping of every fixture, and should prefer, if they were obtainable, as they no doubt will be at a future day, self-cleans-

---

* For a discussion of plumbing appliances see the author's books in a foot note on page 37, particularly his "Domestic Sanitary Appliances."

ing and seal-retaining traps, placed as close to the outlet of fixtures as possible, and made partly or entirely of glass, with the water seal fully exposed.

I should make arrangements to secure to each fixture in a building an ample and never-failing supply of water. In the case of water closets, slop hoppers and urinals, I should always use a separate flushing cistern for each fixture or group of fixtures, while as regards the other fixtures I should give preference to those arranged and constructed in a manner so as to constitute in themselves a small flush tank, thus securing, by their quick discharge, (through outlets larger than commonly in use) a thorough cleansing and scouring of the waste pipe serving them.*

I should, finally, never have a direct connection between any water cistern, a refrigerator or ice chest in a house and the drains or soil pipes, and I should guard with particular care the purity of the supply for drinking purposes.

---

* Fixtures having this important advantage are described in the author's work on "Domestic Sanitary Appliances."

I have, as far as the space at my disposal permits, outlined the leading requirements of a proper system of house drainage, and I confidently assert, from practical knowledge and experience, that, wherever they are conscientiously followed, satisfactory results cannot fail to be secured.

A few points, however, have not been referred to. There is, for instance, the "trap on the main drain" question, which is still agitating the minds of many. I do not feel inclined to be dogmatic about it, for in my own practice I have never followed an iron rule, but have, on the contrary, in each special case, carefully considered and weighed the circumstances and conditions affecting the question. I always use a trap on the line of the drain, if the latter discharges into a cesspool or any form of tank in which sewage is stored for some length of time. I generally advise the use of a trap where the house drain connects with a foul sewer, as in the majority of cases in our large cities. If I use a trap, I should insist upon having it easy of access (but not so as to be exposed to freezing), and provided with proper cleaning hand-holes. Where a house drains into a street sewer, forming part of a well-planned general system of well-flushed sewers, ventilated by open soil pipes in the houses, constructed under supervision of a competent engineer, I should not object to the omission of the trap, always provided the work in the house is thoroughly well done. Where the owner would not mind the additional expense, I should probably prefer to arrange for the ventilation of the sewer by having a pipe carried up to the roof, along the outside of the house, thus preserving a complete disconnection of the house interior from the sewer. Whenever I use a trap, I should also arrange a fresh air inlet, to induce a current through the soil pipe system. I should, however, strongly advise my clients not to terminate the inlet in a box on the sidewalk, covered with a grating, as is now so often done, for such a grating frequently becomes obstructed and closed in winter time. Nor should I carry the fresh air pipe up to the roof.

Where to arrange the inlet is a matter which can be determined only in each special case, and which ought never to be restricted by a hard and fast rule, often entirely defeating the purpose for which the inlet was established.

The question of material most suitable for drains, soil pipes, waste and supply pipes, has not been alluded to, nor have I, in my above recommendations, referred to the many kinds of water seal and mechanical traps advocated to exclude sewer air from the fixtures. I do not feel disposed to enter into a discussion of the merits and disadvantages of the "venting" of traps, the object of which is largely to prevent siphonage. Experiments have established with a sufficient degree of certainty, the fact that self-cleansing S-, P-, or running traps, cannot be depended upon always to retain their water seal against siphonage, unless air is admitted at the crown and sewer side of the trap, either by some anti-siphoning trap attachment, or by a so-called "back-air" pipe, of ample size. Consequently, I should not use such traps without providing such protection as the remedies mentioned afford. Later experiments have shown that an air pipe is not a reliable protection against siphonage in *all* cases, especially where the course of the air pipe is long and tortuous, and that where fixtures are not in constant use, it furthers the evaporation of the water in traps, and hence endangers the safety of plumbing work. That it increases the cost of plumbing, and hinders simplicity of arrangement, must be conceded by all. Thus, while it offers certain advantages in some instances, there are other cases where the disadvantages predominate. It remains then to be decided, only after a thorough and intelligent consideration of all conditions, whether a seal retaining water seal trap, safe against back pressure, siphonage or other influence, or an anti-siphoning trap attachment of some kind, may not be preferable. The question cannot, in my judgment, be decided in an off-hand way. Being of grave moment for the safety of the inmates of a house, the question of trapping should receive an earnest, thoughtful and unprejudiced consideration. Unfortunately, discussions on this point, in sanitary and architectural journals, have not always been divested of useless and much to be deprecated personalities.

As with traps, I also prefer to omit lengthy descriptions of any special plumbing fixtures. There are now a number of each kind in the market, possessing merits and giving satisfaction, which a judicious house owner may select, guided by the above hints.

In conclusion, it must always be borne in mind, that no system of plumbing or drainage will work forever, without proper care, attention and periodical inspection; that stagnation of water or air must be avoided in drain, soil and vent pipes, as well as in traps; that the water in the latter should be frequently changed; and that, in the *tight jointing, safe trapping* and *constant ventilation of pipes*, together with the *frequent flushing* and *thorough cleansing of fixtures*, consist the principal safeguards of a proper system of house drainage and plumbing against entrance and diffusion of noxious sewer air.

III.
THE DRAINAGE
OF A
HOUSE.

THE DRAINAGE OF A HOUSE.*

In the "Homes of To-day" no feature is, to my mind deserving of more attention from architects and house-builders than the sanitary arrangements, yet this very feature, which conduces so much to the well-being, comfort, and happiness of the occupants of a dwelling, is the one to which, until quite recently, far too little importance was attached. I believe I am not mistaken when I assert that the drainage of a house is, probably, to most architects, still the least attractive part of the numerous details of house construction. This fact is not surprising if we remember that a true architect should, above all, be an artist. Men who combine depth of artistic feeling with a profound knowledge of methods of construction and the principles of sanitation, as applied to house-

---

*This paper appeared originally in the "*Homes of To-day*," published by Frank L. Smith, Boston, Mass.

building, are rare indeed. Hence a new profession has sprung into existence, the members of which began to devote their attention to the hitherto neglected branches of architecture, not merely to the drainage and sewerage, but also to the ventilation, heating, lighting, water-supply, and much else involved in dwelling-house sanitation. Mr. Robert Rawlinson, C. E., has well said that "sanitary engineering is a new science, and as its main purpose is to make health, comfort and a prolongation of life practicable, its study to a useful purpose must be important."

A large part of the writer's professional work consists in the proper arrangement of the sanitary drainage of buildings, hence it is assumed that the following brief statement of some of the more important facts concerning house drainage will be of general interest.

During the last decade much progress has been made in sanitary knowledge, and in particular in the art of draining houses, and not the least useful result accomplished has been the better education of the general public in the details of domestic sanitary matters. Twenty or more years ago householders cared little or nothing about the final disposal of the foul wastes from houses. They were content if the plumbing work was arranged so that a free flow and discharge of water could be obtained at each sink, tub, or basin in the house, and appliances of improper construction, from a sanitary point of view, were retained, from ignorance or from reasons of false economy. Noisome and disagreeable odors about a water-closet were often tolerated as being necessary accompaniments of such fixtures. The danger of exposure, night and day, year in and year out, in bed-rooms, living-rooms, or offices, to an atmosphere polluted by gases resulting from the decomposition of stagnant sewage matters, was wholly ignored, and the warnings of early reformers generally disregarded. In city dwellings the ample supply of water, which in turn serves as a vehicle for transporting refuse matter, and the more general introduction of the convenient plumbing fixtures, led, owing to the leaky condition of brick or earthenware drains under houses, to a sewage-sodden condition of the soil under basements. This is true not only of the vast number of buildings erected by shrewd speculators, but it applies alike to the palatial mansions of the rich.

Indeed, the death-rate from zymotic diseases increased, not only in houses with damp cellars, basement, and foundation-walls, but principally in those elaborately planned and richly furnished residences of the better class, where innumerable stationary wash-bowls, defective in arrangement, and tightly enclosed by decorative cabinet work, were scattered in bed-rooms all over the house. As the chief faults of the plumbing work in such dwellings, the following may be enumerated, viz.: the unnecessary multiplication of fixtures, with its accompanying complication of the work; the leaky joints of soil and waste pipes; the broken and leaky drains; the coating of soapy or greasy slime attaching to the walls of waste-pipes; the partial or utter absence of ventilation; furthermore, the defective methods of trapping; the untrapped openings for the drainage of cellar floors leading to the house-sewer; the accumulation of grease in traps under kitchen and pantry sinks; the lack of flushing in all parts of the pipe system, resulting in an accumulation of putrefying slime; the concealment of all work, and the bad workmanship of hidden parts of the plumbing; the untidiness of the spaces under fixtures; the injudicious location of water-closets and bath-rooms, and in particular the faulty position of the closet for servants' use, in out-of-the-way corners, without light and air; lastly, the befouled condition of servants' closets and housemaids' sinks, the offensiveness of the hidden interior of objectionable pan-closets, the deficient water-supply at fixtures on upper floors of city houses, the inefficient flush of valve-closets, the insufficient strength and unreliable support of lead pipes, and the careless exposure of plumbing work to injury by frost.

To say that all this has been changed in the past years would hardly be true, but it is safe to assert that radical improvements have been carried out. In some cities the most urgent reforms are now enforced by law, at least in the case of new houses. Yet, notwithstanding all this, it must be said that the character of the plumbing work in most modern houses is susceptible of much improvement, as I hope to be able to demonstrate. In my own practice, my chief aim has always been to awaken an interest in simplicity of construction, and, in this respect, my practice may differ from that of other reformers. I have, from time to time, made statements, describing what, in my opinion, are the cardinal requirements of good house drainage,* and I claim for them merely that they are, first, the outgrowth of a large and varied practical experience in the supervision of drainage works in new houses, and in the remodelling of defective work; and, secondly, that they are the result of a careful study and comparison of all the sound methods proposed for the improvement of the sanitary condition of our homes.

---

* See the various works of the author.

Before alluding to the cardinal principles and fundamental requirements of good drainage, let me give a few words of general advice to people who intend to build. To begin with, if you build a house, keep the plumbing and drainage as a separate matter from your house contract. It cannot be denied that, where the whole work is given to one contractor, his chief interest—often his only interest—lies in the prospect of pecuniary gain. Thus, as a rule, the plumbing work is sub-let by him as cheaply as possible. There are, of course, among builders good men; but the result is, in at least nine cases out of ten, that the owner pays more to get an inferior job, and —what is more serious—in a house built for his own occupancy endangers the health of the members of his family by exposing them to the minor disorders of the system, to the graver ailments, and to the sometimes fatal diseases associated with bad drainage. To get even tolerably good work under the circumstances, is certainly the exception rather than the rule.

Supposing then, that the owner follows this part of my advice by keeping the plumbing separate, the next question is, whether plumbing work should be contracted for at a stated sum, or whether it should be done by day's work. I have, some years ago, pronounced emphatically against drainage work done by contract, and I see no reason now for changing my opinion. I still believe that the fairest way is to have such work done on a fixed percentage of profit to the contractor on all labor and material. At the same time I cannot deny, and have frequently demonstrated in my own practice, that an entirely satisfactory plumbing job may be obtained by contract work. In this case, however, it is absolutely necessary that the fixtures be properly located, the work carefully planned and arranged on scientific principles, that the contract be based upon a strict and detailed specification, and that the work be placed under intelligent supervision. Even then it is a wise precaution to hesitate to award the contract to "the lowest bidder," universal as the habit may be. It is an axiom of all good sanitarians, which the general public has been slow in accepting, that no house should be occupied as a human habitation until its sanitary condition, as regards drainage, sewerage, ventilation, and kindred matters, has been thoroughly tested. Therefore, it is evident, that, in building a new house, much subsequent trouble and annoyance, not to mention serious illness, may be avoided if the above details of sanitary construction are put at once into the hands of an expert.

A few progressive architectural firms now follow the radical departure of employing regularly a sanitary engineer or plumbing expert to look after the sanitary details of houses. A number of others have the moral courage to tell their clients that they much prefer to have a specialist control the plumbing work in houses built under their supervision. I venture to predict that it will not be many years before attention to sanitation will be universally practised, and sanitary construction be represented by specialists in the leading architectural offices. There would certainly seem to be ample work on hand, and the results, so far obtained where this practice is followed, would seem to justify its more general adoption.

At present the prevailing custom is to leave too much of the detail of the work to the discretion of the plumbing contractor. It is quite evident, that, as the plumber's chief interest is that of a business man, he will not make particular efforts to simplify the work submitted to him, by adopting safer and less complicated methods than those called for by the average specification, and by cutting out and dispensing with unnecessary fixtures. Of course there are exceptions, but they are decidedly in the minority. As a rule, plumbers are too apt to sneer at any attempt of a radical departure from the methods of work handed down to them by tradition.

Owners, on the other hand, often nowadays, place too implicit confidence in the supervision carried out by boards of health. While the results accomplished in cities where plumbing work is subject to regulations and official inspections have been most gratifying, it can not and should not be expected that, even with an increased force of inspectors, every individual house will receive sufficient attention. To illustrate: I have in mind a newly built house on the West side of the upper part of the city of New York, which I was recently asked to inspect by a client who had purchased it immediately upon its completion. The work was done under board-of-health supervision, but evidently by a rascally contractor, with the result that the new owner had to spend about seven hundred and fifty dollars to put the plumbing into a merely tolerably good condition, by recaulking fraudulent joints in iron pipe, by re-fitting water-closets left with broken earthen trap connection, by remodelling sinks left imperfectly trapped, and by putting in a proper system of tank-water supply,—the house being fitted with plumbing apparatus on the upper floors, where in daytime the city water supply failed. In order not to be misunderstood, I desire to state expressly that this is not cited as illustrating the imperfect supervision of boards of health,—for their inspectors accomplish as much good as would seem possible under the circumstances, considering the vast extent of the building districts assigned to each of them,—but only to warn the public against putting too much weight upon the statement, now so frequently encountered in announcements of real estate agents and building speculators, that "*the plumbing work was done under supervision of the board of health.*"

Again, neither architects nor owners should allow themselves to be guided—as is unfortunately too often the case— by the advice of dealers in plumbers' supplies or manufacturers of sanitary specialties. No matter how intelligent and ingenious they may be, their judgment cannot be unbiassed. This is so obvious as hardly to require any further explanation.

In matters of drainage, perfect safety lies in absolutely faithful and faultless work. This can only be attained by employing first-class, honest, and thoroughly competent workmen, and by using first-class materials, fittings, and apparatus, by which I do not mean gilt-edged, embossed or decorated bowls, costly cabinet-work, fancy marble-work or tiling, and nickel or silver plated pipes. All such features are only "for show," and a perfect job, from a sanitary point of view, may be secured without them.

To give to housebuilders specific advice, I should counsel them to avoid all complication, and to aim at simplicity; to avoid having plumbing fixtures not in daily or constant use; to have what fixtures are needed conveniently located, without scattering them injudiciously over the house. Too much convenience in the shape of a profusion of fixtures increases the risk; while by reducing the number of openings into the waste pipe system, the amount of piping, and hence chances of leakage, are reduced correspondingly. Plumbing work should be confined mainly to the bathroom, the kitchen, pantry, and laundry. Some well-meaning friends have repeatedly expressed surprise at the stand-point taken by me in advising the banishment of all fixtures, washbowls or others, from sleeping and living rooms. I desire to state distinctly that I consider it entirely feasible and practicable, in the present advanced state of the art of draining houses, to have in each bedroom of a house the luxury of a stationary washstand, with an abundant flow of hot and cold water, and made perfectly secure against entrance of sewer air. Convenient as such "set" basins may be, I, as a rule, advise dispensing with them in sleeping apartments, and unventilated closets adjoining them, in view of the possibility of imperfect work, particularly where plumbing inspection is not insisted upon. Moreover, it should not be overlooked, that, however safe plumbing fixtures may be originally constructed, the possibility remains of their becoming unsafe under careless use and management.

In advising the employment of competent sanitary experts, I have, to some extent, a personal interest in view. This I cannot deny; but the force of the advice is not weakened by this admission, and it should not be overlooked that the public is ultimately the gainer. In support of my argument in favor of expert superintendence as regards sanitary construction, I may be permitted to quote what others have well said before me: "Sensible people, when they are ill, consult a physician, and not an apothecary. When they wish to plan a house they take the advice of an architect, and not a builder. Both apothicary and builder are of course necessary." So it is also with sanitary experts. The sanitary engineer and the plumber are both necessary; but, while the execution of the drainage works of a house should be intrusted to a plumber, the design of the drainage system should be in the hands of a disinterested engineer. That so many householders, although considering the plumber "the pillager of their purses," still should persist in relying in the majority of cases solely upon his advice, is a matter beyond comprehension to me.

Whole volumes may be, and have been, written describing and explaining the general principles of sanitary plumbing. The essential points may be summed up in the following brief rules, viz.:—

Avoidance of any retention of filth on the premises, by complete, automatic, and instant removal of all waste matter before decomposition takes place.

Thorough ventilation of the whole drainage system.

Abundant and frequent flushing of all fixtures, traps, and waste pipes.

Secure trapping of all vessels having openings in communication with the waste pipe system.

Avoidance of all manner of mechanical obstruction to the flow of waste water.

Durability of the work, soundness of materials, and tightness of joints.

Perfect accessibility to all parts of the work.

Noiselessness in operation of all fixtures.

Prevention of unnecessary water waste by leakage, by freezing, or during flushing.

The cardinal rule in planning should be to observe the greatest possible simplicity of arrangement consistent with convenience and comfort. A fundamental requirement is the reduction of the number of fixtures, and another the concentration of waste discharges through as few well-ventilated pipe channels as possible. As an instance from my own practice I may mention that in re-arranging the plumbing work for the main building of a large insane asylum in this State, I have grouped not less than ten water-closets, thirteen basins, three slop-hoppers, seven bath-tubs, one urinal, and one sink, on four floors, around a single line of soil pipe, kept freely open at the top and at its lower end; thus gaining not only the advantage of greatly reduced cost, but the benefit of an abundant flushing of the only soil-pipe, together with compactness of arrangement. A multiplication of soil pipe stacks and long lateral waste pipes must both be avoided. Each fixture should have a direct and short connection to the soil pipe, if possible by a separate Y-branch. Each fixture should be separately and securely trapped. Where long branch waste pipes are unavoidable they should have separate independent vent pipes through the roof. Lateral branches to the soil pipe, if not more than a few feet long, do not need this ventilation, provided the fixtures are quick emptying and in frequent use; for at each discharge of the fixture a movement of air takes place sufficient to avoid stagnation. All soil pipes should have ample ventilation at top and bottom, and their mouths above the roof should be enlarged, and kept unrestricted by any form of cover. All basins, tubs, and sinks should have large waste outlets, to empty quickly and to fill the waste pipes, thereby securing a thorough scouring of the sides. I am in favor of using large supply pipes and valves and faucets with free waterway; but I also recommend using small waste pipes and small traps, as having a greater tendency to keep clean. Outlets of water-closets, on the contrary, should in my judgment be restricted as much as is consistent with their safe use. Every discharge vessel in a house should act as a flush tank. All fixtures should be of a strong, durable, non-absorbent, and non-corrosive material, with smooth surface, and free from corners favoring accumulation of foulness. The question of overflow pipes has been solved in a simple manner by the introduction of a number of excellent appliances, doing away entirely with concealed overflow passages of any kind.

I favor the entire exposure of all spaces under plumbing fixtures and about pipes, for it should be constantly borne in mind that even the best workmanship and material cannot be expected to last forever, and leaks or other defects are more readily detected if the above advice is followed. Any one who has had occasion to carry out sanitary inspections will bear me out in the statement that all inspections are greatly facilitated where work is kept exposed. It should, therefore, be laid down as a rule in new work, to keep every thing in sight, to leave pipes and fixtures exposed to view, and traps and stop cocks accessible. This also promotes cleanliness, and greatly facilitates the carrying out of needed repairs or alterations.

I advise closing all free communication, by the pipe channels, betweeen the various floors of a building, in order to avoid the carrying about of local odors from one part of a house to another.

Without making any attempt to be exhaustive, I will mention at least a few matters of construction. All soil and waste pipes should be of heavy iron pipe, and restricted in diameter, so as to increase the flushing effect of a stream passing through them, thus avoiding deposits and subsequent stoppages. All piping should be made both air and water tight, drains should be laid with proper fall and true alignment, junctions should be made with Y-branches, and cleaning hand-holes should be provided in places where needed.

Traps for fixtures should have no enlargements or corners favoring accumulations of slime or sediment, and no mechanical obstructions should be countenanced. Traps should be self-scouring, made readily accessible by tight-fitting yet easily removable clean-out caps, and should have a water seal of sufficient depth, and perfectly secure against self-siphonage, back-pressure, capillary attraction, siphonage, and evaporation. From my best knowledge and belief, I cannot accept as universally necessary the requirement of "back ventilation" of traps. I conform to it, as a matter of course, wherever local board-of-health regulations require it; as I have also been compelled—always under protest—to run fresh-air inlet-pipes to grated openings in sidewalks, which choke with ice and snow in winter-time, and to cover soil pipes with the objectionable return bends and vent caps. I do not fail to explain to my clients that the back airing of traps is done at the expense of simplicity; that, in a properly laid-out system, trap vent pipes are not necessary to prevent dead ends in short lateral waste pipes; and that prevention of siphonage can be accomplished, and the extra cost incurred by back-air pipes be saved, in all but rare instances, by adopting simpler and well-known devices. Where I am compelled to run back-air pipes, complicating the pipe system, it is always my endeavor to modify the arrangement, so as not to expose the water in the trap too much to the air current; for there can be no question that the thereby increased free circulation of air in the vicinity of the sealing water of traps hastens the unsealing, by evaporation, of traps under fixtures which remain unused for some days in succession, and endangers the security of all traps during any period when a house is left unoccupied.

Water-closets have now come into almost universal use, even in cottages of moderate cost, and their advantages and comforts over more primitive devices are undeniable. A water-closet is the most important plumbing fixture in the house, and hence should be selected and put up with particular care. A good apparatus should fulfill the following requirements, viz.: it should be simple, neat, and compact in design and construction; durable, strong, and not liable to breakage by careless use; of a smooth material, with ample standing water in the bowl; all parts exposed to fouling should be thoroughly scoured; the flush of the closet should be powerful, quick, copious yet noiseless; the water-closet should be securely trapped, and the trap kept, if possible, accessible and its water-seal visible; it should be free from all machinery liable to get out of order, and should be economical in the use of flushing water required to keep it in a clean condition. There are a number of excellent water-closet appliances now in the market, which practically fulfill nearly all of these requirements.

Properly arranged water-closets will also serve the purpose of a good urinal, and thus do away with a former abomination in houses. A clean slop-hopper or house-maid's sink on the bedroom floor of a house is an undeniable convenience to servants, yet, rather than put it in a dark and unventilated closet, and leave it without means for flushing, I should advise using the water-closet in its stead.

As regards lavatories of all kinds, the first requirement is that the inlet and outlet openings should not be one and the same; for, if so, in filling the vessel, some of the dirty water comes back with the clean. This same objection applies to a number of waste-valves for bath-tubs and basins designed to take the place of the ordinary chain and plug arrangement. The latter device has also radical defects, which are beginning to be more universally recognized and admitted. The outlets of the ordinary chain and plug fixtures are altogether too small in proportion to the diameter of their trap and waste pipe, with the inevitable result that both remain imperfectly flushed, and accumulate to some degree foulness. The chain and plug in the bottom of fixtures is inconvenient in use, and foul slime attaches to the numerous links of the chain, which are difficult to clean. Finally it becomes necessary with such fixtures to use hidden overflow channels, the inside walls of which receive no constant flow of water, hence become readily fouled; and, being arranged so that they cannot be reached, they offer no chance for cleaning. The decaying soap slime coating the overflow passages remains in open communication with the apartment, and forms a serious objection, a standing nuisance, and a menace to health. Numerous patent contrivances have been invented, in which the aim has been to do away with the use of chain and plug, but which retain other objectionable features. Fortunately, wash-basins, bath-tubs, and sinks may now be obtained with standpipe overflow, which answer all the requirements which can be made to such apparatus. They have large outlets, causing a rapid discharge, and securing the incidental advantage of a thorough flushing of the trap and waste pipe. The inside of these fixtures presents a smooth and unbroken surface, the lift devices for the standpipe are convenient in use, and the standpipe itself can be readily disconnected for cleaning purposes. There is less labor for the plumber in fitting up such basins or baths; the number of joints to be made is reduced to a minimum; and every essential part of the fixture, including its discharge and overflow arrangement, is visible and completely accessible.

Concerning tests, as applied to plumbing work during construction and after completion, there cannot any more at this date be the slightest doubt that security for work properly done lies in the clause of the contract specifying that all work will be submitted to rigid tests before being finally accepted. Experience deduced from my own practice is that a better class of work is turned out where these conditions are insisted upon, and I find that mechanics doing first-class work have no objection to any reasonable test applied to their work.

The fundamental rules and requirements hastily sketched above are applicable to all classes of buildings, to dwellings of moderate cost, as well as to mansions and palatial residences. While writing this article I have, among other work, charge of the drainage of a hospital, a schoolhouse, a club-house, a mission-house, a large fire-proof hotel, a row of apartment-houses, and a number of city and country residences, some elaborate and expensive, others of moderate cost and plain design. In all these buildings, without exception, the above leading requirements are being carried out.

There are, of course, numerous other points in the plumbing work of a building requiring attention, such as the arrangement of the supply pipes, the hot-water service, questions of tank supply, pumps and pumping engines, fitting up of hot-water boilers, etc.; but these cannot be discussed without stepping outside of the subject indicated in the title of this essay.

In arranging the drainage of a house, the ultimate disposal of the sewage—a matter usually outside the province of the architect—should not be lost sight of. The sanitary expert must consider the various methods of accomplishing this disposal of the household waste without offence. Where a discharge into sewers or open water courses is inadmissible or unavailable, the disposal of the sewage on the premises is the only alternative. Where sufficient ground cannot be had, a perfectly tight and well-ventilated cesspool, situated at a safe distance from the dwelling, and widely apart from any well furnishing drinking water, is the only device to be tolerated, although it is at best a makeshift involving the temporary storage of noxious and decomposing organic matter.

In the case of suburban and country residences having ample grounds about them, a perfectly safe solution of the difficulty may be found in the adoption of the sub-surface irrigation system, with automatic flush tank, in which the sewage is intermittently distributed under the soil by a net-work of drain tiles laid close under the surface.*

---

*See the author's book in the Science Series entitled, "*The Disposal of Household Wastes*," published by the D. Van Nostrand Co., 23 Murray Street, New York City, 1890.

IV.

MAXIMS
OF
PLUMBING
AND
HOUSE DRAINAGE.

The following notes on plumbing will show more particularly what requirements should be made in regard to materials, workmanship and arrangement of the plumbing work in houses. They should be looked upon and used merely as a framework upon which a specification, suited to each special case, is to be constructed. They are intended primarily to suggest points that require careful consideration, but they also constitute instructions for the guidance of the mechanics engaged in such work.

MAXIMS

OF

PLUMBING AND HOUSE DRAINAGE.

To obtain a safe and secure system of drainage and plumbing in a building, to secure to a house immunity at all times from sewer air, and to prevent any subsequent annoyances incident to bad arrangement and careless workmanship in water pipes and plumbing appliances, it is necessary to observe the following points:

First, to have the system of fixtures, traps, supply and waste pipes well planned and arranged in accordance with the best rules.

Second, to have the work itself constructed in a thoroughly able and efficient manner by competent mechanics and under proper superintendence.

Finally, after the work is all completed and put to use, to have it managed with intelligence, properly taken care of, and examined from time to time, as to its continued soundness and freedom from defects, as is done with all other mechanical apparatus and machinery.

Hence, the subject naturally divides itself into the following sections, viz.:

A. Principles which should govern the planning and location of plumbing work in dwellings.

B. Rules regarding the proper construction of the work, in particular as to

(1) *Materials.*
(2) *Workmanship.*
(3) *General Arrangement.*
(4) *Tests.*

C. Suggestions as to the management and proper care of plumbing apparatus.

It is well known that much of the success of a system of house drainage and plumbing, to say nothing of the convenient and compact grouping of fixtures and the thereby reduced cost of the work, depends upon a judicious planning and arrangement by the architect or person designing the plans of the building.

The rules regarding the proper construction of the work will contain the principal requirements which, in whole or in part, should be embodied in plumbing specifications for all kinds of buildings. It is, however, absolutely necessary, in order to make such a specification complete and adapted to any particular building or dwelling house, that in addition to general requirements the specification should contain a detailed enumeration of all plumbing appliances (fixtures, traps, supply and waste pipes, tanks, boilers, flushing cisterns, stop-cocks, faucets, etc.) required in the building; a detailed and accurate description of the location of the plumbing fixtures, and of the special apparatus wanted, and a minute description of the course of all hot and cold water service pipes, soil, drain, waste, vent, overflow and drip pipes. In every case floor plans and all needed sections of the building should be added, showing clearly the proposed system of drainage and water supply.

From a long practical experience with such details of interior finish in newly erected buildings, I am convinced that it is quite important, in order to secure good results, to prepare a plumbing specification with scrupulous regard to details and with much thoroughness and exactness. It is a great mistake on the part of architects or owners to suppose that such rules and regulations, or printed blank specifications, as the boards of health in our large cities now require to be filed, are sufficiently detailed to provent gross carelessness or deception on the part of unscrupulous plumbing contractors. Plumbing regulations, as framed by boards of health, are good in their way, although undoubtedly susceptible of much improvement. They are altogether too complicated, too detailed in many respects, while not strict enough in others. All those which have come to my knowledge permit certain things which every sanitary engineer worthy of the name absolutely condemns, and, on the other hand, they specify or require things, some of doubtful utility, and others absolutely objectionable, and which no one who has impartially investigated the subject can conscientiously approve.

The suggestions named under (C) refer especially to householders and servants. It has become a recognized fact that a properly constructed drainage system of a house must be intelligently used, and needs constant care and attention on the part of the householder to maintain it in good order. Especially is this true of the vast number of houses occupied only during a part of the year, such as summer residences, summer hotels, seaside and mountain cottages, etc. It also refers to city houses, many of which are closed and vacated during the hot summer months. How to leave plumbing work in such houses during winter and during summer, without incurring the risk of finding, on returning, that the pipes and traps have been frozen and burst, or that the rooms, walls, carpets and furniture are saturated with sewer air, is a question to which a certain amount of attention might with advantage be paid by every householder.

A. Rules regarding the planning and location of plumbing work in dwellings.

Avoid a useless multiplication of plumbing fixtures. Let the amount of plumbing work in a house be reduced as much as possible. Above all, avoid locating fixtures in unoccupied or spare rooms.

Do not place plumbing fixtures of any kind in sleeping rooms, nor even in unventilated closets adjoining them.

Plumbing fixtures, especially water closets, urinals and slop hoppers, must always be located in well-lighted and well-ventilated apartments.

Always arrange fixtures so as to be concentrated, as much as is consistent with convenience in use, in compact groups. Have as few vertical lines of pipe as possible. Avoid long horizontal runs of pipe.

Arrange all plumbing work in a simple manner, with as little complication as is attainable.

In small cottages place the bath room as nearly as possible over the kitchen, in order to reduce the amount of piping, and to simplifiy the whole work.

In small houses it is preferable to separate the water closet from the bath room and to give to each of them a well-lighted and ventilated apartment.

In houses with several toilet and dressing rooms adjoining the bed rooms, the water closet, lavatory and bath-tub may, however, be arranged together.

Avoid locating any fixtures, such as laundry tubs or servants' water closets, in the cellar.

Servants' water closets, slophoppers and housemaids' sinks should never be located in dark, out-of-the-way corners.

Avoid locating any supply pipes on outside walls, especially if the house stands detached and exposed.

In country houses occupied during the

winter, do not locate the bath room and the water tank in that corner of the building which is most exposed to the weather and the direction of the coldest wind (generally the northwest corner).

Place all soil, waste and supply pipes outside of walls or partitions. Let pipes pass in sight through closets, and leave them fully exposed in bath rooms.

Avoid dead ends in all except short branch waste pipes.

In larger houses arrange means for drawing hot and cold water on every floor. Provide a flushing-rim slop-hopper on bedroom floors for the convenience of the servants.

Avoid having plumbing apparatus on floors where you are not sure of a constant, abundant and never-failing supply of water.

Openings in the cellar floor, connecting to a house sewer, should be avoided as unsafe, even if properly trapped.

Common overflow pipes and waste pipes of fixtures which are not in daily use are objectionable.

Arrange all fixtures, wherever possible, so as to have distinctly independent outlets into the main soil pipe system.

Quick-emptying fixtures, constituting a small flush tank for the waste pipe attached to them, and arranged without concealed overflow channels, are preferable to other kinds.

Select plumbing fixtures of a strong, durable, smooth, non-absorbent and non-corrosive material, of a simple construction, and with as little movable mechanism as possible.

Arrange all fixtures in an open manner. Avoid carpentry of any kind, enclosing plumbing apparatus, as much as possible.

Avoid lead "safes." Floors and walls about plumbing fixtures should preferably be made water-tight, and covered with slate, marble or encaustic tiles.

Wherever much grease is wasted, provide a suitable grease trap.

Keep, as a guide in case of future examinations, plans showing the location of all drain pipes, traps, access holes, stop-cocks, etc.

B. Rules regarding the proper construction of the work.

In the following are given the principal requirements, which, in whole or in part, should be embodied in plumbing specifications for all kinds of buildings. The subject is subdivided under the headings; (1) General conditions; (2) Materials; (3) Workmanship; (4) General Arrangement of the Work; (5) Tests of the Work during Construction and after Completion.

I desire, however, to have it distinctly understood that the following is in no sense intended to be, or to be used as, a general plumbing specification, which can be copied *verbatim*, or which, by filling in lines usually left blank for the convenience of architects, may be readily adapted to any kind of work. The following notes should only be considered a help in preparing a plumbing specification. If intelligently used, I have no doubt they will prove useful to those who are anxious to write a good and complete specification.

1. General Conditions.

All the work contemplated, shown on the floor plans and in the sections of the building, and described in the specification, shall be done in the best and most workmanlike manner, to the satisfaction of the superintendent and of the owner.

The plumber must furnish all material and perform all labor required to finish the work contemplated in the specification in a substantial manner.

He must do all his work promptly as the building progresses, and must in particular not delay other contractors nor interfere with their work.

The contractor is not to sub-let the whole or a part of his work without the written approval of the owner or his representative.

The plumber must lay out his own work correctly according to the floor plans, and is to give his personal superintendence to the work.

The superintendent shall have access to the work at all times, and shall be sole judge of the quality and fitness of the materials used, and of the character of the workmanship.

No pipe, fitting, or work of any kind to be closed up or hidden from view before it has been examined and approved by the superintendent. Any unfaithful or imperfect work, or defective material that he or the owner may discover before the work is finally accepted, shall be immediately corrected, and any pipe, fitting, trap, fixture or material of any kind which in the superintendent's judgment does not conform with the requirements of the specification, shall be at once removed and replaced at the contractor's expense by satisfactory work and material.

The plumber shall be guided in his work by both drawings and specifications. Preference must in all cases be given to figures or memoranda, and only where these are not given, scale measurements may be taken. Wherever the specification varies or conflicts with the drawings, the plumber is to be governed by the specification.

The plumber is to obtain all official

permits required, pay the fees for the same, and is to give to the proper authorities all notices required by law relating to his work.

All work must conform with the local building and health regulations, and the latter are to be considered a part of the specification.

The plumber must see to it that no damage is done to any part of his own or other contractors' work on the building. He will be held responsible for all soiling of walls, wainscots, finger marks or other defacements by his workmen. He will see that proper care is taken in kindling fires in the plumber's furnace, and in handling the latter anywhere in the building.*

The plumber must see to it that all building rubbish caused by his operations be removed from time to time from the building as may be required. At the completion of the work he is to deliver everything in a clean condition and in good working order, and perfect in all respects.

---

* A good clause to insert is the following: "No smoking or spitting allowed in building after the plastering is done, the trim set and floors laid."

The contractor will be paid only on certificates properly signed by the superintendent.

[Here should follow a schedule of the fixtures required in the building.]

2. Materials.

All the materials used in the work to be of the best quality obtainable in the market.

Earthenware Drain Pipes.

Outside drains (beginning at a distance of at least five feet from the house) to be of strong, salt-glazed earthen pipes, either pipes provided with hubs at one end, or else plain cylindrical pipes with loose rings or collars of unglazed earthenware.

All vitrified pipes to be perfectly straight, circular and true in section, of a uniform thickness of not less than three-quarter inch for four and six-inch pipes, to be free from cracks, flaws, or other defects, to be hard-burnt, not brittle, smooth and impervious on the inside and highly glazed, except at the pipe ends. Hubs of vitrified socket pipes to be not less than three inches deep.

Earthenware special fittings, such as T and Y branches, bends, offsets, traps, etc., to be of the same quality and character as specified for pipes.*

Cast-iron Drain, Soil, Waste, Vent, and Leader Pipes and Fittings.

Cast-iron drain, soil, waste, vent and leader pipes to be of a homogeneous texture, free from flaws, cracks, sand-holes or similar defects, perfectly straight, truly cylindrical, perfectly smooth on the inside and of a uniform thickness of not less than one-quarter inch. Pipes to be the ordinary bell and spigot joints, with hubs of great depth and strength of metal.

Pipes to be tested and inspected at foundry, and to be afterwards coated and

---

* See also " Specification for Laying House Drains and Pipe Sewers in the author's book, " *The Disposal of Household Wastes* " No. 97 of Van Nostrand's Science Series, Published by D. Van Nostrand Co. , New York. 1890.

protected inside and outside with coal tar pitch or other equivalent substance. Or else, pipes to be dipped in metallic paint, or to be enameled with black enamel, or to be porcelain-lined (white enamel), as the detailed specifications may require. The superintendent may, instead, require that the pipes and fittings be tested in his presence by the oil of kerosene test.

All fittings for cast-iron soil and waste pipes to be of best quality cast iron; all castings to be sound, clean, smooth, true, free from flaws, cracks, sand-holes, air bubbles or other imperfections or impurities in the metal. Thickness of shell to be not less than one-quarter inch. Fittings for bell and spigot pipes to have hubs which must be very deep and extra strong. All fittings to be tar coated or enameled, to correspond with the kind of pipe required by the detailed specifications.

Cast-iron soil pipes to be four inches inside diameter, and iron waste pipes to be two inches diameter, unless other sizes are specifically called for. Drain pipes to be from two to six inches diameter, as called for in detailed specifications, and as indicated on the plans.

Lead Pipes.

All lead pipes to be drawn pipes of soft pure lead, of the best make. Pipes to be either plain or tin-lined, as may be directed in detailed specification.

Weight of lead pipe to be as follows:

a. Supply Pipes.

Inside Diameter.	$\tfrac{3}{8}''$	$\tfrac{1}{2}''$	$\tfrac{5}{8}''$	$\tfrac{3}{4}''$	$1''$	$1\tfrac{1}{4}''$	$1\tfrac{1}{2}''$
			Lbs. per running ft.				
For street or tank pressure not exceeding 20 lbs. per square inch.........	1	$1\tfrac{3}{4}$	$2\tfrac{1}{2}$	3	4	5	$6\tfrac{1}{2}$
For street or tank pressure exceeding 20 lbs. per sq. in., and not exceeding 35 lbs. per square inch.........	$1\tfrac{1}{2}$	2	3	4	$4\tfrac{3}{4}$	6	8
For pressures exceeding 35 lbs. per square inch	2	3	4	5	6	7	9

b. Waste, vent, overflow and drip pipes, also supply pipes from w. cl. cisterns to w. cl. bowls.

½ inch	lead	pipe	to	weigh	1	lb.	per running ft.
¾ "	"	"	"	"	1¾	"	"
1 "	"	"	"	"	2	"	"
1¼ "	"	"	"	"	2½	"	"
1½ "	"	"	"	"	3½	"	"
2 "	"	"	"	"	5	"	"
3 "	"	"	"	"	6	"	"
4 "	"	"	"	"	8	"	"

Block Tin Pipes.

Suction pipes in wells or cisterns to be of pure block tin, to weigh as follows:

Lbs. per running foot.

¾ inch	block tin	pipe	to weigh	¾
1 "	"	"		1
1¼ "	"	"		1¼
1½ "	"	"		1½

Brass Pipes.

All brass piping to be made of seamless drawn brass tubing, with all the necessary brass fittings required. Pipes to be either plain (for hot water) or lined inside with tin (for cold water). Outside of pipes to be nickel plated and polished, or else, if brass finished, pipes to be varnished with a good coat of shellac, as may be required in detailed specifications.

Wrought-iron Pipes.

a. Supply Pipes to be either plain, galvanized, enameled, lined inside with block tin, or rubber lined, or made rustless by the Bower-Barff process, as may be directed in the detailed specifications. Pipes to be uniform and true in section and pipes in sizes up to 1¼ inches to be butt-welded and warranted to be tested by hydraulic pressure of 300 pounds per square inch. All necessary fittings used to have the same protecting treatment as the pipes. All fittings to be malleable iron fittings.

b. Soil, Waste and Vent Pipes to be of standard wrought-iron pipe, having a uniform thickness of not less than one-quarter inch; pipes to be lap-welded and proved at the iron mills to 500 pounds per square inch by hydraulic pressure; to be coated after being heated, with a preparation of coal tar and asphalt (or to be treated with the Bower-Barff or other rustless process.) Fittings for soil and waste pipes to be protected against rust by the same process as applied to the pipes ; to be tapped truly straight and to have a strong shoulder.

Traps.

Traps to be of lead, brass, copper or glass, for sinks, basins and tubs, and to be be of iron, lead, stoneware or earthenware under water closets (as may be specified in describing the water closet apparatus.)

Traps of lead not to be lighter in weight and thickness of lead than the waste pipes to which they are attached. Drawn lead traps to be preferred to cast lead traps; the latter, if used, to be proved free from sand-holes, flaws or other imperfections. Traps for water closets to have at least one inch, and—except in the case of certain special water closet constructions — not more than two inches water seal. All other traps for fixtures to have at least one and one-half inch effective seal.

All traps to be self-cleansing and free from corners or spaces favoring accumulation of filth, and to be of such shape as to hold as little volume of water as is consistent with a proper water seal.

Traps under fixtures so be provided with cleansing trap screws, placed below the water level in the traps, and arranged so as to be accessible.

No bell or D-trap, or cesspool trap, to be used anywhere (special kind of trap required to be stated in the detailed specifition).

Traps for house drains, leaders, yard and area drains to be of heavy cast-iron provided with proper tightly closed cleaning hand-holes and to have a proper depth of seal of not less than two inches.

Stop-cocks, Valves and Faucets.

Roundway stop-cocks to be used on main lines of service pipes, and to be of the same inner diameter as their respective pipes. If supply pipes are of wrought iron, brass full way gate valves, instead of stop-cocks, to be used. All valves, faucets and stop-cocks to be of superior gun metal of the heaviest existing patterns. No globe valves to be used.

Plated Ware.

Wherever brass is to be plated the nickel or silver plating (as specified in detailed specification) to be of the best kind, and warranted for at least five years.

Solder.

All solder used on the work to be pure, soft, and free from impurities, such as zinc or other admixture.

Pig Lead.

Lead for caulking purposes to be soft and pure pig lead, to be free from any admixture of antimony, zinc or other metal. No lead used for caulking soil pipes to be brittle or hard; no scrap lead to be used.

Sheet Lead.

To be of soft and pure pig lead, rolled

in sheets, and to be free from any admixture of alloys or old scrap lead. Weight to be not less than four pounds per square foot.

Sheet Copper.

All sheet copper to weigh sixteen ounces per square foot, unless greater weight is called for in detailed specifications, and to be well planished and tinned.

Putty.

The use of putty is to be avoided, but, wherever needed, putty to be made up with pure linseed oil, and to be mixed with some red lead, to avoid its being eaten by rats.

Marble.

All marble is to be of best Italian quality, blue veined and highly polished, unless light pink Tennessee or other marble is especially called for in the description of the plumbing fixtures. All slabs are to be properly molded and countersunk. All marble work to be sharply molded and well fitted and set, and, where necessary, to be secured with round-head brass screws and washers, the latter sunk in flush.

Fixtures.

All fixtures and apparatus to be strictly of the kind and character and special make as called for in the following detailed description, and every fixture to be free from any defects, perfect throughout, and put in complete working order.

[Here should follow a complete description of the special appliances wanted under the head of

Hydrant; sillcock; furnace or steam boiler supply.
Temporary cock in cellar to furnish water for building purposes.
Double-acting force pump; pumping engine.
Water tank.
Kitchen range; kitchen boiler; kitchen sink.

Cistern pump.
Wash trays; laundry sink; laundry range; clothes boiler.
Refrigerator; ice sink.
Water closets.
Wash-basins.
Bath-tubs, foot-baths, sitz-bath, bidets, shower-baths, needle-baths.
Slop hopper; housemaid's sink.
Urinal and ladies' urinette.]

Cement.

Hydraulic cement to be any good, pure, quick-setting brand, to be fine and freshly ground; cement to be subject to inspection and testing by the superintendent, who may reject all cement of improper quality.

Sand.

To be clean, sharp, silicious sand, free from all dirt, dust loam or other foreign matter.

Mortar.

Sand and cement to be mixed dry and to be wetted up only in small quantities as used, and with just enough water to make a paste of proper consistency. Mortar to consist of one part hydraulic cement and two parts of sand—measured quantities. No lime to be used in the mixture, nor should any mortar be used that has begun to set.

Concrete.

Concrete for foundations and trench bottoms to be composed of one part mortar and five parts broken stone. The mortar to be made up as directed above, with sufficient water only to constitute a fair paste. The broken stone shall then be wetted up and mixed with the paste in a thorough manner by several turnings with a shovel. The concrete, after being thrown in place, to be tamped with wooden rammers. No concrete left over at the close of a day to be used afterwards on the work.

3. WORKMANSHIP.

Earthenware Drain Pipes.

If plain cylindrical pipes are specified, joints to be made by means of loose collars or rings set in mortar. Drain pipes to be laid at a depth of generally not less than three feet, in a carefully excavated trench. Trench to be opened only as wide as necessary and to be suitably braced wherever required, to prevent any caving in of the sides. Trench to have the bottom trimmed perfectly to the exact grade, and to have at each pipe joint a depression, in order to have each length of pipe evenly and perfectly supported. Pipes should be laid with ends butting close together. Cement mortar, mixed as per directions given, to be applied to the unglazed ends of the pipes, and also to the inside of collars. Pipes to be joined in such a manner that the flow line will be true and even, in order to prevent stoppages. The inside of each length of pipe to be well cleaned before laying down the next length. Drains to be laid in perfectly straight lines, and deviations from alignment to be made with special curves of large radius. Each length of pipe to be covered in the center with a few inches of earth, in order to steady it and to prevent any pipes from moving. After testing pipe joints with water, the back filling to be done with great care, in order not to disturb the pipes. Earth to be thrown in layers of not more than twelve inches in depth, and the filling to be well rammed or puddled, to prevent the slightest settling. All outside branch drains to join the main by Y, not T branches. Bends in the lines of the drains should never be made with straight pipes. Pipes of different sizes to be joined by proper reducing fittings. Long lines of house drains to have manholes or inspection holes placed at suitable intervals.

If earthen socket pipes are specified, sockets to be examined with special care for cracks or flaws, before lowering pipes into the trench. Trench to be excavated as described heretofore. Special grooves to be cut in the bottom of the trench for the sockets, in order to give to the pipes a firm bearing throughout their entire length. Pipes to be laid with the socket pointing up grade. The space between spigot and hub, if the hub is deep, to be filled first with a small gasket of oakum, to prevent any cement mortar from entering at the joints. The remaining space to be filled with hydraulic mortar, which must be applied with particular care at the bottom of the joint. Water accumulating in the hollow grooves must be removed before applying mortar to the joint. Some cement to be wiped at the face of each joint, and as soon as the joint is finished, the grooves to be filled with earth, in order to support the cement at the joint until the cement has had time to harden. The utmost care to be observed after a joint is made, to prevent any disturbance of the pipes by stepping on them, or otherwise. The inside of each pipe joint to be thoroughly cleaned from any projecting oakum or cement. Back filling to be done as described heretofore.

Earthen pipe must always be laid on a firm bed, which should be provided in case of loose soils. A bed of gravel or sand, or a concrete foundation, should be prepared in such cases, to properly support the pipes.

Joints between earthen pipes and iron house drains to be made perfectly tight by means of pure hydraulic cement.*

Cast-iron Pipes.

All joints in cast-iron socket pipe and fittings to be made by inserting a gasket of picked oakum into the space between spigot and hub, so as to fill about one-half of the depth of hub, and pouring molten soft lead from a large ladle into the remaining space. After cooling and shrinking, the lead to be thoroughly caulked with caulking tools to insure air and water-tight joints. The face of the caulking lead to remain without paint, putty or cement, so as to leave the marks of the caulking tool exposed to view. All joints made before soil pipe and fittings are put in place must

---

* For a more detailed specification, the reader is referred to the author's book, "The Disposal of Household Wastes,' in the Van Nostrand Science Series.

be re-caulked when the soil or waste pipe stack is up, as they may otherwise become loose from jars.

Iron pipe lengths in a vertical position to be firmly held in place by strong, wrought iron staples of round iron, or to be supported by strong pipe rests or supports, well-fastened to the walls and placed under each hub. Where hung from the ceiling, pipes to be held in place by improved iron pipe hangers, securely fastened to joists, and drain pipes carried above the floor, to be supported at suitable intervals by brick piers. No ordinary plumber's hooks to be used.

Suitable provision to be made, especially in very high buildings, for the free contraction and expansion of each vertical stack, by fastening the pipes in such a manner that a slight movement parallel to the pipe axis may take place, especially where a waste pipe receives much hot water.

All changes in direction to be made with bends of large radius, and all branches to be Y or half Y branches, or curved Tees (the latter only on upright lines).

After all pipes are put in place they are to be cleaned and painted, so as to look neat at the completion of the work.

Lead Supply and Waste Pipes.

All joints between lead pipes, whether for supply or waste pipes, to be wiped solder joints; cup joints not to be made anywhere.

Joints between lead pipes and brass fittings, such as stop-cocks, ferrules, etc., to be wiped solder joints; joints between lead pipes and brass couplings may be cup joints.

Joints between lead pipes and cast-iron pipes or hub fittings, to be made by means of a brass or copper ferrule, soldered to the lead pipe and tightly caulked into the iron hub.

Joints between lead pipes and wrought-iron pipes, or tapped fittings, to be made by means of brass male and female soldering nipples, wiped to the lead pipe, and tightly screwed with a wrench into or onto the tapped fitting or threaded pipe.

Connections between earthenware and lead to be made, as much as possible, by means of brass couplings.

All vertical lead pipes to be supported by hard metal tacks, placed at short and equal distances, and securely fastened by screws to finished boards put up by the carpenter.

Horizontal or graded lead pipes to be firmly supported throughout their whole length on boards, to prevent sagging and trapping, and to be fastened and kept in place by brass bands (strips of sheet brass bent to conform to the shape of the pipe) placed at frequent intervals and fastened to the boards by screws. Hot-water pipes to be fastened, preferably by brass bands only, not by tacks, so as to allow for necessary expansion and contraction. Hooks should never be used in fastening lead pipes.

Joints in tin-lined lead pipes to be made by means of the special tinned brass ferrules and fittings furnished with the pipe by its manufacturers. If wiped solder joints are made, unusual care is required, so that, in applying the heated solder, the tin may not be melted and thus the pipe lining be destroyed.

Wrought-iron Pipes.

Joints in wrought-iron pipes to be made with screw threads cut on the ends of the pipes and in the shoulder of fittings. All threads to be of standard gauge. All cut ends of pipe to have the burr removed. A thick paste of red and white lead mixed, or else linseed oil, to be used in the joints to act as lubricant, and to make up for imperfections in the threads. Pipes to be screwed together tightly with wrenches or pipe tongs, care being taken that no lead is squeezed out at the inside of pipes. In putting together wrought-iron pipes and fittings treated with black enamel, a liquid black enamel should be used at the joints, to protect the threads against rust.

Wrought-iron pipes, lined inside with tin, should be joined by means of the special ferrules sold for such purpose with the pipe.

Horizontal wrought-iron pipes to be supported by means of special pipe hangers. All exposed vertical lines of iron supply pipes are to be carried perfectly plumb and straight at an even distance from the walls, and secured with galvanized iron or brass holdfasts arranged so that they can be readily removed.

Brass Pipes.

All brass pipes and brass fittings to be put together with screw joints, a paste of red lead being put over the threads, and the joints made perfectly tight by means of screw wrenches. Brass pipes to be put up on boards and fastened by brass bands, holdfasts and escutcheons, or to be supported by neat brass hangers, varnished or nickel-plated. No hooks to be used. Long horizontal lines of brass service pipes for hot water should not be confined at ends, but should be arranged so as to allow of free expansion and contraction.

Stop-Cocks and Faucets.

Joints between bibbs or stop-cocks and lead pipes to be wiped solder joints; those between bibbs or stop-cocks and wrought-iron or brass pipes to be screw joints.

Traps.

All joints between lead traps and lead waste pipes to be wiped solder joints, made perfectly tight. Joints between lead traps and cast-iron soil pipes or fittings to be made with brass ferrules, soldered to the lead pipe and caulked into iron hubs. Junctions between trap and waste pipe to be made perfectly tight with the greatest care. Junction between fixture and trap may be made movable, to facilitate repairs of the fixture.

Water-closets having traps located above the floor must have joints at the floor with the soil pipe made with particular care by means of brass floor flanges, in the case of earthen, and by means of caulked joints in the case of iron traps, so as to be perfectly tight.

Water closet traps of iron or lead below the floor must have their weight well supported, to prevent the joint at the floor from being torn loose and becoming leaky.

Traps under fixtures, leader traps, and the trap on the main drain must be set perfectly true as regards their water level. All bending or tipping of lead traps set between floor joists must be avoided. In all cases traps are to be placed as close to fixtures as possible.

Fixtures.

Workmanship in fixtures to be of neat appearance throughout, whether work is to remain exposed or not. All fixtures, with the appliances belonging to them, to be properly set in good and complete working order.

In General.

All plumbing work, whether to be left exposed to view or to be boxed up, to be done in a thorough manner.

Putty joints to be avoided wherever feasible.

All openings into the pipe system, as well as fixtures set in place, must be securely covered up, to prevent obstruction of the pipes or breakage of fixtures by carelessly or maliciously dropped materials.

The plumber must cut no beams, joists, floors or studs; this will be done for him by the carpenter. Plumber is to arrange all pipes necessarily placed between joists so as to run, wherever possible, parallel to the direction of the beams. Pipes placed between joists to be run with proper grade, and to be continuously supported on sound boards nailed over strips tacked to the joists.

Where pipes pass through the roof, an absolutely water-tight joint to be made around pipes. If flashings are used for this purpose they should be at least eighteen inches square, of heavy sheet lead or copper, with a funnel slipped over the pipe, turned over and caulked into a hub, while the ends of the flashing are tightly fastened to the roof.

Wherever plumbers' pipes pass through floors, ceilings, walls or partitions, the plumber must see to it that the holes are neatly and perfectly closed around the pipes in the floor, and that the holes in the ceiling be closed up and well plastered. If required, sleeves of galvanized iron pipe are to be used in the case of supply pipes, with neatly-fitted polished and lacquered or nickel-plated brass escutcheons screwed on at the floor and ceiling, or either side of the walls, as the case may be.

4. GENERAL ARRANGEMENT OF PLUMBING WORK.

The whole work to be arranged and executed in strict conformance with the specifications and the floor plans and sections, showing the plumbing work and in particular the *exact location* of every plumbing fixture in the building. Unless special permission is obtained from the superintendent to deviate from the lines as laid out in the drawings, these must be strictly followed.

[Here insert a detailed description of the course of all soil and waste pipes, vent and supply pipes.]·

Soil and Waste Pipes.

All main lines of drain, soil, waste and air pipes inside of the building to be of heavy iron; short branch wastes and vents from fixtures and traps, and branch supply pipes to be of heavy lead pipe.

All soil and waste pipes and supply pipes to be placed where shown on plans, and their whole arrangement to be as compact and direct as possible.

Each vertical stack of soil or waste pipe to run as straight as possible, avoiding offsets, up to the roof, and to be continued to a point at least three feet above the same, so as to have the mouth well exposed to air currents. Extensions above the roof to be at least full size in the case of all soil pipes; it is preferable, however, to enlarge soil pipes to six inches above the roof, and waste pipes should be enlarged to four inches before passing through the roof. None of the pipes above the roof should be smaller than four inches, because smaller openings are liable to clog and freeze up in winter time.

Mouths of all soil, vent and waste pipes to be kept at a safe distance from ventilating shafts, dormer windows, skylights or chimney flues. Vertical pipes run along chimney flues to terminate at least two feet below the top of the flue.

Mouths of all pipes above the roof to be kept *wide open*. No return bend, ventilating cap or patent ventilator to be used. Where obstructions are anticipated, carry the pipe at least six feet high, and where this is impracticable, cover the mouth of the pipe with wire gauze, or insert a mushroom-shaped wire basket.

No soil or vent pipe to be run to and to terminate in any hot or cold flue or ventilating shaft.

Soil pipes receiving wastes from water closets to be four inches in diameter, all other waste pipes to be two inches. No deviation to be made from these sizes unless ordered or approved by the superintendent. Each soil and waste pipe stack to have proper fittings to receive branch wastes from fixtures.

No soil, waste or vent pipe to be connected with any chimney flue. No soil or waste pipe to be used to carry rain water. No trap to be placed at the foot of any vertical stack of soil or waste pipe.

Junction between Vertical Pipes and Main Drain.

Junction between vertical stacks of soil or waste pipe and main drain to be made with Y branches and eighth bends, or, at the upper end, with bends of easy sweep. Junctions to be supported by strong brick piers.

Main Drain.

Main drain in cellar to be kept above the floor, *in sight*, unless otherwise directed by the superintendent. Size of main drain to be not less than four nor more than six inches in diameter, unless specially ordered by the superintendent.

Grade of pipes in cellar or basement to be not less than one-quarter of an inch nor more than one inch to the foot, unless by special approval of the superintendent. All branches to join the main with Y-branches pointing in the direction of the flow.

Cleaning Hand-holes.

Cleaning hand-holes (closed tightly by trap screws) to be provided near all junctions between vertical and horizontal pipes, and at junctions of horizontal branch pipes with the main drain, also near bends and traps.

Trap on Main Drain.

Main drain to be trapped (unless the trap is to be omitted by special order of the superintendent, in which case the fresh air inlet can be dispensed with) where it leaves the house walls by a running or by a ½ S-trap of iron, with proper cleaning hand-holes, arranged accessible, but not exposed to freezing. Over the main trap and all other cleaning hand-holes, if below the cellar floor, arrange cast-iron frames, set in masonry or cement, and having chequered iron covers.

Fresh-air Inlet.

From just inside the trap run a 4-inch fresh-air inlet, terminating at a point above the surface, well remote from windows, and with opening protected against obstructions. (The superintendent will decide exact location of the fresh-air inlet.)

Leader Pipes.

Vertical pipes for the removal of rain water from roofs, placed inside of the building, must be of cast-iron or wrought-iron with tight joints. No waste water from any plumbing fixture to deliver into any leader pipe.

Trapping of Leaders.

Outside leaders of metal (galvanized iron, copper, tin), with slip or soldered joints, and also leader pipes of whatever material, whether located inside or outside the building, must be trapped in case the top opens below or near windows, or near flues or ventilating shafts. Iron leaders with tight joints, with tops remote from windows, are to be left untrapped.

Traps for leaders to have a seal of more than ordinary depth, to provide against evaporation. Traps for leaders and those for yard and area drains not to be buried out of sight or covered with concrete, but to be in all cases placed where they are protected against the action of frost, in easily accessible positions, and to be provided with cleaning and inspection handholes, with well-fitting and tight-closing covers.

Drainage of Areas and Yards.

Areas, court yards and paved open spaces to be properly drained by trapped branch drains, the trap to be located preferably inside of the cellar walls, protected against freezing. Openings in the yard or area to be covered with well fastened brass strainers or iron gratings, protected against rust. No bell traps to be used.

Waste Pipes for Fixtures.

Each fixture to have a separate and independent connection to the main soil pipe (unless otherwise approved by the superintendent). In no case shall basin or bath tub wastes discharge into a water closet trap below the floor.

Branch wastes from fixtures to be carried as directly as possible to the soil or waste pipe. Branch waste pipes carried under floors to be as short as practicable, and if of lead to have a continuous support, to prevent sagging.

Trapping of Fixtures.

Each fixture connected to the soil or waste-pipe system to be provided, as near as possible to its outlet, with a suitable trap secure against siphonage, back pressure, evaporation, etc. [The kind of trap should be specified in the detailed description of each apparatus.]

No fixtures to be provided with more than one trap. No trap under a fixture to be of larger bore than the waste pipe to which it is attached.

All traps under fixtures to be arranged so as to be readily accessible, and to be provided with cleaning hand-holes or trap-screws, located below the water-line of the trap.

Round pipe traps of the S, half S, or running shape, not to be used unless provided with a ventilating pipe, or some other effective attachment, to prevent siphonage.

Trap Vent Pipes.

Wherever vent pipes are used, the branch vent pipes for water-closet traps should be not less than two inches in diameter. All other traps to have vents of same area as the trap. The size of the main vertical lines of vent pipes will depend upon the height of the building and should also increase with the number of branches which they receive.

Where back air pipes are carried through the roof, they must be enlarged to four inches, to prevent clogging in winter time in cold climates. All horizontal air pipes must be so graded as to discharge the water from condensation into a trap or waste pipe. T-branches on upright vent lines must always be set at such a height above floor that the branch vent cannot act as an overflow pipe in case the waste should be stopped up.

Size of Waste Pipes.

Waste pipes for fixtures to be in size as follows :

	Inches inside diameter.
For wash bowls	$1\frac{1}{4}$–$1\frac{1}{2}$
For bath tubs	$1\frac{1}{2}$–2
For pantry sinks	$1\frac{1}{4}$–$1\frac{1}{2}$
For kitchen sinks	$1\frac{1}{2}$–2
For laundry tubs	$1\frac{1}{2}$–2
For slop sinks	2–3
For urinals	$1\frac{1}{4}$–2
For a row of basins, tubs or urinals	2–3

No deviation from these sizes permitted unless specially ordered by the superintendent. The weight of these pipes to be such as called for under "Materials."

Overflow Pipes.

Overflow pipes from fixtures must connect with waste pipes on the inlet side of traps, or they must enter the trap below the water line or, finally, they may be arranged similar to safe waste pipes. They should be entirely avoided wherever pos sible, and hence fixtures without hidden overflow pipes are much to be preferred.

Strainers.

Outlets of all set fixtures, except water closets, to have fixed steam metal strainers, to guard against obstructions.

Safes and Drip Pipes.

All safes, where required under fixtures, to be of of 4-lb. sheet lead, with edges turned up at least two inches all around ; to have a convex brass strainer, well soldered, and a 1-inch drip pipe of lead or rustless wrought-iron pipe carried to a point where a discharge from leakage or otherwise is readily detected. If run to cellar celling, arrangements to be provided to exclude cellar air from the drip pipe. The drip pipe may empty over a sink or cistern, but always so that the discharge may be in sight. In no case should drip pipes be connected with a soil or waste pipe.

In most cases safes and drip pipes may safely be omitted if the work is well done, and all fixtures set in an open manner.

Flushing Cisterns.

Each water closet, urinal or slop hopper should be supplied with water from a special copper or lead lined flushing cistern. The pipe, from the cistern to the fixture, must never be less than $1\frac{1}{4}$ inches in diameter and should be run from the cistern to the bowl as directly and straight as possible. All ballcocks in flushing cisterns must be regulated so as to work noiselessly and without spattering.

Refrigerator Wastes.

Waste pipes from refrigerators or ice-chests to be trapped, and not to have a direct connection with any drain, soil or waste pipe. Arrangements for cleaning and flushing these pipes must be provided.

Tank Overflow.

Overflow pipes from tanks must not discharge into any soil, drain or waste pipe. They must be run into the roof gutter, or else discharge over a sink in the basement, or be carried and emptied into the nearest fixture where the discharge will be visible.

Open Arrangement of Fixtures and Pipes.

All fixtures to be arranged in an open manner, unless otherwise directed by the superintendent.

All soil, waste, vent, supply or drip pipes to be kept exposed to view, or to be cased in woodwork, fastened with screws so that the pipes may remain readily accessible. All piping to be kept outside of partitions, unless otherwise ordered by the superintendent. No pipes to run between floors and ceilings unless *absolutely* necessary.

All spaces about soil, waste or supply pipes, where these pass through floors and ceilings, to be closed absolutely tight in a neat and substantial manner.

Arrangement of Supply Pipes.

The whole arrangement of supply pipes to be as compact as possible.

[Here insert a detailed description of the water supply for the proposed building.]

All supply pipes to be kept outside of floors, walls and partitions, being left exposed and in full view, unless specially otherwise directed by the superintendent. All exposed iron pipes are to be neatly bronzed, if required, with silver or gold bronze, and varnished.

No water supply pipes to run on outside walls, nor to be placed in any position where they would be liable to freeze, unless absolutely necessary, and in this case pipes to be securely protected in exposed places by some non-conducting material, as may be required by the superintendent.

All horizontal lines of supply pipes to be arranged neatly, and laid out so that they will not cross each other or dip one under the other. Supply pipes not to have any depressions or sags, nor to be bent up in their course to avoid their becoming air bound, and causing an interruption in the circulation in the case of hot water pipes.

All supply pipes to be so graded and arranged that they may be easily and completely emptied.

No check valves to be used on any supply pipes unless specially called for in the specifications.

Hot and cold water pipes to be kept at least one-half inch apart everywhere.

To prevent injury to decorated walls and ceilings from drippings arising from condensation in warm weather along cold water pipes, especially if of iron, pipes should be carried across floors in safes made of zinc, and provided with a drip pipe run to the cellar sink.

Size of Supply Pipes.

Branch supply pipes to fixtures to have the following sizes, unless otherwise directed:

For wash bowls	$\frac{1}{2}$ inch bore.
For bath tubs	$\frac{3}{4}$ "
For pantry sinks	$\frac{5}{8}$ "
For kitchen sinks	$\frac{3}{4}$ "
For laundry tubs	$\frac{5}{8}$ "
For slop hoppers (to draw water)	$\frac{3}{4}$ "
For flushing cisterns	$\frac{1}{2}$ "
For flushing pipes from cisterns to water closets, urinals or slop hoppers	$1\frac{1}{4}$–$1\frac{1}{2}$ "

For weight of pipes, see under "Material."
Where a branch pipe supplies more than one fixture, it should be increased in sectional area proportionately.

Rising main to be at least ¾ inch in size; direct branches from it to be of the same size, and pipe from tank to boiler to be not less than ¾ inch. Connections between water back in range and boiler to be made with ¾ inch (better 1-inch) brass or stout brazed copper pipes.

Hot Water Supply.

Hot water boilers, wherever practicable, to be supplied from a tank in the attic, not from street pressure. Main hot water pipe must always be extended from above the highest fixture full size to the top of the tank, where it should be turned over to allow steam to escape; also to prevent the collapse of the boiler.

Stop-cocks.

Stop-cocks for both the hot and cold water supply to be arranged near each fixture (also near each flushing cistern), to shut off the water separately from each fixture if required.

All branch supply pipes to be arranged

so as to be shut off separately by stopcocks or gate valves, and, if required, to be arranged so that they may, each separately, be completely drained.

All stop-cocks on supply pipes to be arranged easy of access.

Faucets, especially ground-key and self-closing bibbs, not to be placed at the end of a line of supply pipe, but to be taken from the side of the pipe, and the pipe to be continued so as to form a small air chamber.

5. TESTS OF THE WORK DURING CONSTRUCTION AND AFTER COMPLETION.

Test of Earthen House Sewer.

Before refilling the trenches for outside drains the earthen sewer pipe and its joints to be tested by closing the main outlet and filling the sewer with water so as to have a pressure corresponding to at least two feet head of water at its upper end, and all joints to be proven tight to the satisfaction of the superintendent.

Test of Pipe System inside the House.

After the completion of all the piping in the house, and before any fixtures are connected, the tightness of joints and soundness of pipes to be tested. All openings of waste, soil and vent pipes and the outer end of house sewer to be securely closed, and the whole system of piping to be filled with water, which must remain at the same level for at least 12 hours. In winter time other tests — smoke test, peppermint or fumes of sulphur test, pressure test with force pump and manometer—to be substituted for the water pressure test. If any of these tests reveal a leak the defect is to to be made good, and pipes will again be tested until the system is proved gas and water-tight to the satisfaction of the superintendent.

Test of Supply Pipes.

All iron and brass supply pipes are to be tested with pressure pump and mercury gauge, and all defective pipes and fittings removed and replaced by sound material, and all leaky joints made tight.

Final Test of the Completed Work.

The whole plumbing work is to be tested after completion by turning the water into the pipes, fixtures and traps everywhere, in order to detect imperfect joints or bad pipes, or holes caused by careless driving of nails.

The whole system is, finally, to be tested in the presence of the superintendent by the oil of peppermint test, or fumes of burning sulphur, introduced by means of an "asphyxiator." Any defects found to be at once repaired by the plumber, who is to bear the whole expense, and all to be left in perfect working order and warranted for —— years.*

C. **Rules regarding the proper Care and Management of Plumbing Apparatus.**

Even the best sanitary appliances, discharging quickly through self-cleansing

---
* Very often the gas-fitting work is included in the plumbing specification. For a description of gas fitting work, the reader is referred to the author's book in D. Van Nostrand's Science Series, entitled, *Notes on Gas Lighting and Gas Fitting, with a Specification for Gas Piping and Some Hints to Gas Consumers.*

traps and well ventilated and abundantly flushed waste pipes, need constant care and frequent cleaning.

Plumbing fixtures, and all traps, soil, drain and waste pipes, require periodical inspection, same as a steam boiler or other machinery. In order to be readily inspected they should be kept accessible. Therefore avoid all enclosure of the plumbing work.

The whole security of plumbing work lies in thorough workmanship, good materials, safe trapping, abundant flushing, constant ventilation and *absolute purity.*

The water in traps under any kind of plumbing fixture must be frequently changed.

Stagnation of water or air should be avoided, not only in the drains and vent pipes, but in traps as well.

A judicious use of the fixtures and proper cleanliness are indispensable to keep plumbing apparatus in a sweet and wholesome condition.

Water closets and slop hoppers in particular, but other plumbing fixtures not to any less extent, should be thoroughly cleaned and scrubbed with soap, hot water and a scrubbing brush, at least once a week, and as much oftener as possible.

The same care and treatment should be applied to the floors and walls surrounding the closet, and to the woodwork of the seat. Hence it is important, in order to facilitate cleaning operations, to arrange all plumbing work in an open manner.

Even where fixtures are cased up with ornamental woodwork, let the parts be readily removable; avoid nailed carpentry, and never allow any accumulation of rags or rubbish of any kind under the water closets, basins or sinks.

After cleansing the sides of bath and laundry tubs and wash basins, let clean water from the faucets run for some time into the fixtures, in order to change completely the water standing in the trap.

After pouring out slop jars or pails into slop hoppers, always flush the fixture and its trap by one or more discharges from the flushing cistern.

If you have plumbing work in spare

rooms, closets or guests' bedrooms, let some one of the household make it a daily practice to turn on the water, to make sure that the traps are constantly filled.

If much grease is emptied through kitchen or pantry sinks, it is advisable to rinse occasionally the waste pipes and traps by pouring through them a hot and concentrated solution of potash.

In leaving a city house for the summer months, when evaporation of water in traps is most active, *especially with vented traps*, all overflow holes in wash bowls, pantry sinks and bath tubs should be closed by corks, or by pasting paper over the openings, then close the outlets with plugs and fill basins and tubs with water to near the overflow line. In the case of kitchen sinks it is best to remove the open strainer, substituting a plug strainer, closing the outlet with a plug and filling the sink with water. Wash-tubs may be similarly protected by closing the outlets and filling the tubs with water. Fixtures without overflow pipes (with stand pipe outlets) are more easily protected than those in common use, and are preferable on this as well as on many other accounts. In case of slop hoppers and water closets, it becomes necessary to dip out all water from the trap and to fill the trap with glycerine or oil, or a solution of chloride of calcium. Water closets and slop hoppers flushed from automatic siphon or tilting tanks may continue to receive the flush at intervals, provided their branch supply pipe is taken out in such a manner that it will not interfere with the shutting off of the water in the remaining parts of the house. Trap attachments may also be had which continue to keep the trap filled with water up to the proper water level, if evaporation or loss by siphonage or capillary attraction should take place.

In leaving a country residence for the winter, it is of the utmost importance to remove completely all water from all supply and waste pipes, traps, fixtures and cisterns, so that nothing can freeze. Hence it is very necessary in the case of country houses to run all pipes with such a continuous grade that they may be completely drained and emptied. The water supply should be thoroughly shut off in the basement or cellar, taking care to open all faucets and stop-cocks at fixtures. The kitchen boiler must be completely emptied by means of the sediment cock, and also the water tank in the attic. Next remove the water from all water closet and slop hopper cisterns. Traps under fixtures may be emptied by means of the brass trap screws usually provided, at the lowest point of the trap, or by removing the cleanout caps, or else by using a sponge. All overflows must be closed, the traps filled with glycerine, and the outlets of fixtures closed with plugs as previously described. Water closet traps should be filled with a strong salt solution, to which may be added some calcium chloride. As an additional security the trap may be boxed up, and the box filled with sawdust.

V.
MEMORANDA
ON THE
COST
OF
PLUMBING WORK.

MEMORANDA

ON

THE COST OF PLUMBING WORK.

While this volume was going through the press it occurred to me that a few notes on the cost of work, of such a character as is described in the preceding pages, and as is now required in the best examples of drainage and plumbing of buildings, would be of particular usefulness to architects, to their clients, and to all people contemplating the remodeling of their plumbing work.

What moves me particularly to publish some memoranda regarding the approximate expense of such work is the fact that there seem still to exist in many quarters very vague ideas on this point. The majority of people who build houses will insist upon having numerous and elaborate plumbing appliances. Generally, however, on receiving bids for the work they are disagreeably surprised about the "exorbitant" figures asked for. Such people should remember that plumbing and drainage work of the *best* character will cost more than the flimsy, unsanitary work put, until recently, into most houses, and in particular into those erected by unscrupulous contractors, or by the cheap or speculative builders. It is useless and wrong to make careless statements, such as the following, which I find in a recent architectural publication:

"In an ordinary household, numbering six or eight persons, occupying an average dwelling, there will usually be a single bath room, two water closets, one up stairs, and one for the use of domestics, a kitchen sink and hot water boiler, wash trays, butler's pantry sink, and from one to five stationary wash basins, while if the water pressure is deficient in the city, or if in the country the roof water is stored for family consumption, a tank or cistern will be required. In such a dwelling a four-inch soil pipe will be ample for the drainage from the principal fixtures, and a two-inch cast-iron waste pipe from the basins, if any are located at a distance from the main lines. There will also be need for a two-inch waste from the kitchen fixtures, together with a five-inch rain leader, and connections from surface cesspools in the front and back yards. All these pipes will discharge into a five-inch main drain leading to the sewer or cesspool. *To plumb a house of this grade will cost in the neighborhood of three hundred dollars.*" (The italics are mine.)

Such statements are apt to do a great deal of injustice and even harm to respectable plumbing contractors who strive to do work of a high character, and do not expect to make more than a legitimate profit on their contracts. It is absurd to expect the work enumerated in the particular example quoted to be done in a proper manner at such a low figure. *Twice the sum named* would, from my judgment and experience, be a moderate and tolerably correct estimate of cost, and even this would suppose the work to be of a plain character, although satisfactory from a sanitary point of view.

The figures given below apply to fixtures, completely put up, with their supply and waste pipes, traps, vents, water fittings, such as faucets and stop-cocks, including all labor on same:

Drain and Soil Pipes (fittings included):
    4" pipe, extra heavy.........$1.00 per ft.
    5" "   "   .... 1.25 "
    6" "   "   .... 1.50 "

Vent and Waste Pipes:
    2" pipe, extra heavy......... .50 "
    3" "   "   .... .75 "

Main Trap with Manhole and Cover......$10.00
Leader Traps, 4"............................ 3.00
                  5"............................ 4.00
Cesspools in Areas.......................... 2.00
Double-acting Force and Lift Pump.... 35.00
Hand or Cistern Pump..................... 10.00
Ericsson Caloric Pumping Engine, 6-inch,
    about..250.00
Rider Pumping Engine, 6-inch, about...450.00

		Gallons.				
Tank, of wood,	100	200	300	400	500	600
copper lined,	$15	$30	$40	$50	$60	$70
Wrought iron, painted,	$50	$65	$75	$85	$95	

Fittings for water tank complete, including ball-cock, stop-cock, overflow, blow-off, cistern valve........................ $20.00

Kitchen Boiler, with water back connections, all necessary couplings, stop-cocks and boiler stand complete

	Gallons.						
	30	40	50	60	70	80	100
Of galv. iron....	$30	$35	$45	$50	$55	$60	$70
Of copper.......	40	50	—	75	—	90	120

Kitchen Sink, of cast-iron. completely fitted up, according to length and pattern, from................................. $15 to $25
Add for galv. and enameling sink.... 10 to 15
Of soapstone, from............ .... 20 to 35
Of slate, from................................. 18 to 30
Of earthenware, from..................... 25 to 40

Butler's Pantry Sink,

Of copper (24 oz.), from................... $20 to $30
Of enameled iron, from.................. 20 to 30
Of steel, painted.......................... 15 to 25
        enameled........................ 25 to 35
Of porcelain, from... .... 30 to 40
Of German silver............ .... 30 to 35
Add for waste valves or stand pipe overflow................................... 10 to 15

Laundry Tubs, each *tub* fitted up completely.

 Of wood... $12
 Of slate... 20
 Of cement..................................... 25
 Of artificial stone........................... 15
 Of soapstone.................................. 30
 Of cast iron (rustless)..................... 18
 Of porcelain, American.................. 38
 Of earthenware, imported.............. 45

For a *set of two* deduct 5 per cent. from twice above sum.

For a *set of three* deduct 10 per cent. from three times above sum.

Water Closets.

 Flushing-rim short hoppers, completely fitted up, excluding wood-work and tiling $50
 Improved washout closets.................... 60
 Improved hopper and siphon closets......... 75

Tiling for water closet walls and floors per square foot, complete, about.................. 1.50

Slate or marble floor slabs, per square foot, about.. 2.00

Water closet seats (without covers)......... $5 to $10

Bidet attachment............................... 5 to 10

Gas jet ventilator attachment............... 10 to 15

Bath tubs, of copper, 16 oz........................ $30
      18 oz.......................... 35
      20 oz.......................... 40

 if fitted with chain and plug.

Add for waste valve or stand pipe overflow.... $15
Enameled iron bath tub............................... 75
Porcelain tub.............................. $180 to $200

Slop Hopper, Improved earthenware, flushing rim slop hopper, with cistern, marble back, etc................... $75 to $100

Urinals. Lipped Bedfordshire urinals, each fitted with flushing cistern...... $30 to $40
Urinal stalls, of slate, for each stall............ 45
       of marble,   "   55

Ladies' Urinette. All porcelain, with flushing cistern, complete..... .... $75 to $90

Ladies' Bidet. Complete................. 60 to 80

Wash Basins. Ordinary wash bowls, chain and plug, round................... $30 to $35
                    oval................. 33 to 38
Improved waste valve or stand pipe basin, round.................................. 40 to 45
                    oval...................................... 45 to 50

Water Supply Pipes (main lines).

For these a suitable sum should be added in the estimate. The amount will depend on sizes and material of supply pipes, size and number of stories of house, etc.

For general and preliminary estimates a rough idea of the cost of the work may be obtained by the following rule : *Count number of fixtures or set of fixtures* (counting the boiler and tank in) *and multiply same* with 50 for ordinary, plain plumbing, iron supply pipes, plain but sanitary fixtures ;

with 60 for very good, but plain plumbing, *i. e.*, best workmanship but plain fixtures ;

with 75 for best quality plumbing materials, lead supply pipes, very best workmanship ;

with 100 for very extensive and elaborate plumbing, including brass hot water pipes, nickel-plated holdfasts, including marble slabs and backs, but excluding all marble slate or tile work for floors and walls and partitions :

with 125 to 150 for most complete work, fitted up with nickel-plated brass piping throughout, and fixtures of the most expensive kind, including marble floor slabs and cabinet-finished, brass-trimmed, or marble-encased cisterns.

The product represents in dollars the approximate cost of the work.

The cost of hot-air pumping engines is, of course, not included, and while the above figure covers cost of all connections between water back and boiler, it does not include the range nor any cabinet work in bath rooms, etc.

Where anti-siphon traps are used under fixtures, and vent pipes are accordingly omitted everywhere, the main system being very amply vented, the cost is reduced from 7 to 10 per cent. from above figures.

VI.

SUGGESTIONS
FOR A
SANITARY CODE.

Suggestions for a Sanitary Code.

A.—Rules as to Healthful Building Construction.

1. It shall be considered unlawful hereafter to erect or cause to be erected a new building upon any site which has been filled up with house refuse or any kind of animal or vegetable matter, unless such matter shall have been properly removed from such site.

2. It shall be considered unlawful hereafter to erect or cause to be erected any new buildings or structures of any kind upon any damp or wet site, unless such site shall have been effectually drained by means of suitable properly laid earthenware tile pipes.

3. It shall be considered unlawful to lay such drain pipes in such a manner as to communicate directly with any drain carrying foul sewage, or with a sewer or cesspool.

4. The drainage of the subsoil of buildings shall conform to the following regulations and requirements:

a. The subsoil drains shall be laid, if possible, at a depth of not less than two feet below the cellar floor.

b. They shall be laid with open joints, protected against entrance of dirt or vermin by paper or muslin wrapping or collars.

c. They shall be laid on a true grade, with perfect alignment and with a continuous fall towards the outfall.

d. The outfall shall be either directly into the open air, or into a ditch or road gutter.

NOTE.—If connection must necessarily be made with a sewer, arrangements shall be made for perfect disconnection, and the water seal of the trap must be maintained, even in the driest seasons, by suitable arrangements, approved by the inspector.

5. Wherever the building site is damp, the cellar floor shall be constructed with at least six inches of concrete. It is recommended to put on top of this a thin coating of coal tar pitch or asphalt, and to

finish it on top with a layer of Portland cement.

It is recommended that every wall of new buildings be provided with a damp-proof course of proper material, placed above the level of the ground, and also that the outside and inside of the foundation wall, to the height of the damp-proof course, be coated with coal tar pitch or asphaltum.

It is recommended to whitewash the cellar walls of all buildings at least twice a year.

6. Buildings without basement or cellar shall be placed on brick or stone piers or posts, and the floor of the first story shall be raised so as to be at least two feet above the surface of the ground. There shall be a free circulation of air underneath the floor, and between it and the surface of the ground.

B.—Rules as to Connection between House Drains and Street Sewers.

1. It shall be considered unlawful to connect or cause to be connected, any

private drain with a street sewer, without first obtaining a permit from the proper authorities.

2. It shall be considered unlawful hereafter to construct any drain for any building and to connect the same to a street sewer, unless the drain shall in its plan and construction conform to the following requirements:

a. Each building shall have a separate connection with the street sewer.

b. Wherever junction pieces have been built into the sewer they must be used for making said connection, unless special permission is obtained to cut the sewer.

c. No pipe or other materials for drains shall be used until they have been examined and approved by the authorities, or their duly appointed superintendent or inspector. No house drain to be larger than five inches inside diameter, except by special permission.

d. No street shall be opened until the

junction piece in the sewer has been located by the superintendent.

e. If no junction pieces are built into the sewer, a connection shall be made by inserting into a brick sewer a junction pipe of proper size, and cut slant to an angle of forty-five degrees by the manufacturer. Great care must be taken not to injure the sewer, and all rubbish shall be carefully removed from its inside.

f. In connecting a house drain with a pipe sewer, a Y junction must be inserted in the line of the sewer, and the main sewer left in a good condition.

g. In all cases the trench must be opened to the point of connection without tunneling, so as to allow of an easy inspection.

h. In opening any street or public way all· materials shall be placed where they will cause the least inconvenience to the public, and the whole inclosed with sufficient barriers, and properly lighted at night from the beginning to the end of the work.

i. The least inclination of the house drain shall be 1 in 60, unless a written permit is obtained to lay the house drain to a lesser grade.

k. When the course of the house drain is not the same as that of the junction piece, it must be connected therewith by a curve of not less than ten feet radius. All changes of direction to be made with curved pipe, and in no case must a pipe be clipped.

l. Every joint shall be laid with gasket and cement, and bedded in hydraulic concrete at least four inches in depth.

m. The ends of all pipes not to be immediately connected shall be securely closed, water-tight, and guarded against entrance of earth with imperishable materials. The inside of every drain, after it is laid, must be left smooth and perfectly clean throughout its entire length, and true in line and grade.

n. The back-filling over drains, after they are laid, shall be puddled or rammed, all water and gas pipes protected from injury or settling, and the surface of the street made good within forty-eight hours after the completion of that part of the drain lying within the public way.

o. No privy vault or cesspool shall be connected with the house drain or sewer.

C.—*Plumbing Regulations.*

1. No plumbing work of any kind shall hereafter be constructed in any building, nor connection made between a house drain and a street sewer, unless said work shall be made to conform strictly to the following requirements:

a. The house drain may be of glazed vitrified pipe with cemented joints to within five feet of the outer line of the house foundation walls. From this point to the inside it shall be of cast-iron pipe, at least one-fourth inch thick, and with joints well caulked with lead, and made air and water-tight.

b. All lines of soil or waste pipes in a building shall be of heavy iron.

c. The house drain shall be trapped, near the point where it leaves the building, by a running or half S-trap, which shall not be larger in diameter than the house drain. This trap shall be placed in an accessible position, protected against freezing, and must be provided with an inspection hole, and a tight closing cover.*

d. The house drain shall not be laid beneath the cellar floor unless absolutely necessary, and in this case it shall be laid in a trench and shall be surrounded with concrete. The trench shall be filled and closed after the

---

* This refers to connections with old and improperly constructed, or foul street sewers, and to cases where house drains discharge into a cesspool or flush tank. For well-constructed, self-cleansing sewers, provided with flushing arrangements and ample ventilation, the trap should be omitted. In the latter instance it should be made a law that in every house connected with the street sewer there shall be an uninterrupted flow of air passing from the sewer up the house drain and soil pipe, and out at the roof, or *vice versa*.

drain is thoroughly inspected and pronounced perfectly tight.

e. All connections in horizontal pipes to be made with Y branches.

f. There shall be a fresh-air inlet pipe, entering the house drain just inside of the main trap, of a diameter of not less than four inches, and opening at any convenient place out of doors, approved by the superintendent or inspector.*

g. All soil and waste pipes shall be run in as straight a manner as possible up to, and at least five feet above, the main house roof. Soil pipes to be enlarged to six inches and waste pipes to four inches above the roof. The upper terminus shall not be located too near a window, ventilating shaft, or chimney flue; the outlet above the roof shall not be capped by either a return bend, ventilating cap, or movable ventilator.

---

* When the trap is not required the fresh-air inlet should be omitted.

h. Extensions of soil or waste pipes shall not be constructed of sheet metal or earthenware, and no soil, waste or vent pipe shall stop in any brick or earthen chimney flue, serving as a ventilator.

i. No soil pipe shall be larger than four inches, and no waste pipe larger than two inches inside diameter (their extensions above the roof excepted.)

k. Before the fixtures are placed in connection with the pipe system, and before the soil pipe and iron house drain are connected with the outside drain, the outlet of the house drain and of all its branches shall be closed tight and the pipe filled with water to its top, and every joint shall be carefully examined for leakage, and all leaks shall be securely closed before connections are made with said pipe system.

l. All soil and waste pipes shall be kept outside of walls or partitions, and the system arranged in such a manner that it may at all times be readily examined and repaired.

m. Every fixture in the house shall be separately and effectually trapped by a seal-retaining trap placed close to the fixture, and arranged so as to be safe against back-pressure, self-siphonage, loss of seal by evaporation or siphonage.

n. No branch waste pipe for tubs, sinks, basins, to be larger than one and one-half inch diameter.

o. Connections of lead pipes with iron hub pipes shall in all cases be made with heavy brass ferrules, properly soldered to the lead, and well caulked to the iron pipe.

p. Every water closet shall be adequately flushed with water from a special flushing cistern arranged directly above it, except that where a cistern is liable to freeze other methods may be permitted, provided that thorough and sufficient flushing is secured. Every water closet apartment shall have direct means of ventilation into the open air. Pan closets shall not be used in any building. The outlets of water closets shall not be larger than three inches in diameter.

q. No opening shall be provided in the house drain for the purpose of receiving the surface drainage of the cellar, unless special permission is previously obtained.

r. All rain water conductors which are carried up within the walls of a building shall be of iron pipe. Connections with such rain water pipes along their vertical course for the discharge of sewage or waste water therein will not be permitted. Rain water conductors shall be trapped if they open at the top near windows, ventilating shafts or flues.

s. It shall be unlawful to throw or deposit, or cause or permit to be thrown or deposited, in any vessel or receptacle connected with a public sewer, any garbage, hair, ashes, fruit or vegetables, peelings, or kitchen refuse of any kind, rags, cotton, cinders, or any other matter or thing whatsoever, except fæces, urine, the necessary closet paper, and liquid house slops.

t. Waste pipes from refrigerators or other receptacles in which provisions are stored, shall not be directly connected with a drain, soil pipe, or other waste or sewer pipes, but shall be made to discharge over an open tray, provided with a waste-pipe and seal-retaining trap.

u. Drip pipes from safes, under any kind of plumbing fixtures, must not have any connection with any soil, waste, or drain pipe.

v. Overflow pipes from water tanks shall not be connected to any soil, waste, or drain pipe.

w. No steam exhaust shall be directly connected with any soil or waste pipe, or drain communicating with a street sewer.

CATALOGUE

OF THE

SCIENTIFIC PUBLICATIONS

OF

D. VAN NOSTRAND COMPANY,

23 MURRAY STREET AND 27 WARREN STREET, N. Y.

A. B. C. CODE. (See Clausen-Thue.)

ABBOTT (A. V.). The Electrical Transmission of Energy. A Manual for the Design of Electrical Circuits. Fourth edition, revised. Fully illustrated. 8vo, cloth........................... net $5 00

ABBOT (Gen'l HENRY L.). The Defence of the Seacoast of the United States. Lectures delivered before the U. S. Naval War College. 8vo, red cloth. 2 00

ADAMS (J. W.). Sewers and Drains for Populous Districts. Embracing Rules and Formulas for the dimensions and construction of works of Sanitary Engineers. Fifth edition. 8vo, cloth. 2 50

A1. CODE. (See Clausen-Thue.)

AIKMAN (C. M., Prof.). Manures and the Principles of Manuring. 8vo, cloth.................... .... 2 50

ALEXANDER (J. H.). Universal Dictionary of Weights and Measures, Ancient and Modern, reduced to the Standards of the United States of America. New edition, enlarged. 8vo, cloth............... ... 3 50

ALEXANDER (S. A.). Broke Down: What Should I Do? A Ready Reference and Key to Locomotive Engineers and Firemen, Round House Machinists, Conductors, Train Hands and Inspectors. With 5 folding plates. 12mo, cloth.................. .... 1 50

ALLEN (C. F.). Tables for Earthwork Computation. 8vo, cloth................................. 1 50

ANDERSON (J. W.). The Prospector's Hand-book; A Guide for the Prospector and Traveller in search of Metal-bearing or other Valuable Minerals. Seventh edition, thoroughly revised and much enlarged. 8vo, cloth.. 1 50

ANDERSON (WILLIAM). On the Conversion of Heat into Work. A Practical Hand-book on Heat-Engines. Third edition. Illustrated. 12mo, cloth. 2 25

ANDES (LOUIS). Vegetable Fats and Oils; their Practical Preparation, Purification and Employment for various purposes. Their Properties, Adulteration and Examination. A Hand-book for Oil Manufacturers and Refiners, Candle, Soap and Lubricating Oil Manufacturers and the Oil and Fat Industry in general. Translated from the German. With 94 illustrations. 8vo, cloth............................ 4 00

—— Animal Fats and Oils. Their Practical Production, Purification and Uses for a great variety of purposes, their Properties, Falsification and Examination. A Hand-book for Manufacturers of Oil and Fat Products, Soap and Candle Makers, Agriculturists, Tanners, etc. Translated by Charles Salter. With 62 illustrations. 8vo, cloth....................net 4 00

ARNOLD (Dr. R.). Ammonia and Ammonium Compounds. A Practical Manual for Manufacturers, Chemists, Gas Engineers and Drysalters. Second edition. 12mo, cloth.............................. 2 00

ARNOLD (E.). Armature Windings of Direct Current Dynamos. Extension and Application of a General Winding Rule. Translated from the original German by Francis B. DeGress, M. E. With numerous illustrations. 8vo, cloth............................ 2 00

ATKINSON (PHILIP). The Elements of Electric Lighting, including Electric Generation, Measurement, Storage, and Distribution. Ninth edition. Fully revised and new matter added. Illustrated. 12mo, cloth.. 1 50

—— The Elements of Dynamic Electricity and Magnetism. Third edition. 120 illustrations. 12mo, cloth.. 2 00

ATKINSON (PHILIP). Power Transmitted by Electricity and its Application by the Electric Motor, including Electric Railway Construction. New edition, thoroughly revised, and much new matter added. Illustrated. 12mo, cloth.... .... 2 00

ATKINSON (Prof. A. A., *Ohio Univ.*). Electrical and Magnetic Calculations, for the use of Electrical Engineers and Artisans, Teachers, Students, and all others interested in the Theory and Application of Electricity and Magnetism. 8vo, cloth, illustr....net 1 50

AUCHINCLOSS (W. S.). Link and Valve Motions Simplified. Illustrated with 29 woodcuts and 20 lithographic plates, together with a Travel Scale, and numerous useful tables, Thirteenth edition, revised. 8vo, cloth................................ 2 00

AXON (W. E. A.). The Mechanic's Friend. A Collection of Receipts and Practical Suggestions. 12mo, cloth 1 50

BACON (F. W.). A Treatise on the Richards, Steam-Engine Indicator, with directions for its use. By Charles T. Porter. Revised, with notes and large additions as developed by American practice; with an appendix containing useful formulæ and rules for engineers. Illustrated. Fourth edition. 12mo, cloth 1 00

BADT (F. B.). New Dynamo Tender's Hand-book. With 140 illustrations. 18mo, cloth.............. 1 00

—— Bell Hangers' Hand-book. With 97 illustrations. Second edition. 18mo, cloth............... .... 1 00

—— Incandescent Wiring Hand-book. With 35 illustrations and five tables. Fifth edition. 18mo, cloth. 1 00

—— Electric Transmission Hand-book. With 22 illustrations and 27 tables. 18mo, cloth................. 1 00

BALE (M. P.). Pumps and Pumping. A Hand-book for Pump Users. 12mo, cloth..... .... 1 50

BARBA (J.). The Use of Steel for Constructive Purposes. Method of Working, Applying, and Testing Plates and Bars With a Preface by A. L Holley, C. E. 12mo, cloth........................ ..... 1 50

BARKER (ARTHUR H.). Graphic Methods of Engine Design. Including a Graphical Treatment of the Balancing of Engines. 12mo, cloth... ..... 1 50

BARNARD (F. A. P.). Report on Machinery and Processes of the Industrial Arts and Apparatus of the Exact Sciences at the Paris Universal Exposition, 1867. 152 illustrations and 8 folding plates. 8vo, cloth... ... . 5 00

BARNARD (JOHN H.). The Naval Militiaman's Guide. Full leather, pocket form... .... 1 25

BARWISE (SIDNEY, M. D., London). The Purification of Sewage. Being a brief account of the Scientific Principles of Sewage Purification and their Practical Application. 12mo, cloth. Illustrated. 2 00

BAUMEISTER (R.). The Cleaning and Sewage of Cities. Adapted from the German with permission of the author. By J. M. Goodell, C. E. Second edition, revised and corrected, together with an additional appendix. 8vo, cloth. Illustrated....... 2 00

BEAUMONT (ROBERT). Color in Woven Design. With 32 Colored Plates and numerous original illustrations. Large 12mo.............................. 7 50

BEAUMONT, W. and DUGALD CLERK. Autocars and Horseless Carriages....... .....(In Press.)

BECKWITH (ARTHUR). Pottery. Observations on the Materials and Manufacture of Terra-Cotta, Stoneware, Fire-Brick, Porcelain, Earthenware, Brick, Majolica, and Encaustic Tiles. 8vo, paper. Second edition.................. .... 6C

BERNTHSEN (A.). A Text-Book of Organic Chemistry. Translated by George M'Gowan, Ph.D. Fourth English edition. Revised and extended by author and translator. Illustrated. 12mo, cloth.......... 2 50

BERTIN (L. E.). Marine Boilers: Their Construction and Working, dealing more especially with Tubulous Boilers. Translated by Leslie S. Robertson, Upward of 250 illustrations. Preface by Sir William White. 8vo, cloth. Illustrated... .... 7 50

BIGGS (C. H. W.). First Principles of Electricity and Magnetism. Being an attempt to provide an Elementary Book for those intending to enter the profession of Electrical Engineering. Second edition. 12mo, cloth. Illustrated.............. .... 2 00

BINNS (CHAS. F.). Manual of Practical Potting. Compiled by Experts. Third edition, revised and enlarged. 8vo, cloth..............................net 7 50

—— Ceramic Technology; being some aspects of Technical Science as applied to Pottery Manufacture. 8vo, cloth.................................net 5 00

BLAKE (W. P.). Report upon the Precious Metals. Being Statistical Notices of the Principal Gold and Silver producing regions of the world, represented at the Paris Universal Exposition. 8vo, cloth....... 2 00

BLAKESLEY (T. H.). Alternating Currents of Electricity. For the use of Students and Engineers. Third edition, enlarged. 12mo, cloth................ 1 50

BLYTH (A. WYNTER, M. R. C. S., F. C. S.). Foods: their Composition and Analysis. A Manual for the use of Analytical Chemists, with an Introductory Essay on the History of Adulterations, with numerous tables and illustrations. New edition.......(In Press)

—— Poisons : their Effects and Detection. A Manual for the use of Analytical Chemists and Experts, with an Introductory Essay on the growth of Modern Toxicology. Third edition, revised and enlarged. 8vo, cloth.. 7 50

BODMER (G. R.). Hydraulic Motors ; Turbines and Pressure Engines, for the use of Engineers, Manufacturers and Students. Second edition, revised and enlarged. With 204 illustrations. 12mo, cloth...... 5 00

BOILEAU (J. T.). A New and Complete Set of Traverse Tables, Showing the Difference of Latitude and Departure of every minute of the Quadrant and to five places of decimals. 8vo, cloth............... 5 00

BOTTONE (S. R.). Electrical Instrument Making for Amateurs. A Practical Hand-book. With 48 illustrations. Fifth edition, revised. 12mo, cloth....... 50

—— Electric Bells, and all about them. A Practical Book for Practical Men. With more than 100 illustrations. 12mo, cloth. Fourth edition, revised and enlarged... 50

—— The Dynamo : How Made and How Used. A Book for Amateurs. Eighth edition. 12mo, cloth... 1 00

BOTTONE (S. R.). Electro Motors: How Made and How Used. A Hand-book for Amateurs and Practical Men. Second edition. 12mo, cloth.............. . 75

BONNEY (G. E.). The Electro-Platers' Hand-book. A Manual for Amateurs and Young Students on Electro-Metallurgy. 60 illustrations, 12mo, cloth.. 1 20

BOW (R. H.). A Treatise on Bracing. With its application to Bridges and other Structures of Wood or Iron. 156 illustrations. 8vo, cloth.................. 1 50

BOWSER (Prof. E. A.). An Elementary Treatise on Analytic Geometry. Embracing Plane Geometry, and an Introduction to Geometry of three Dimensions. 12mo, cloth. Twenty-first edition............ 1 75

—— An Elementary Treatise on the Differential and Integral Calculus. With numerous examples. 12mo, cloth. Seventeenth edition.................. 2 25

—— An Elementary Treatise on Analytic Mechanics. With numerous examples. 12mo, cloth. Fourteenth edition... .... 3 00

—— An Elementary Treatise on Hydro-Mechanics. With numerous examples. 12mo, cloth. Fifth edition... 2 50

—— A Treatise on Roofs and Bridges. With Numerous Exercises. Especially adapted for school use.. 12mo, cloth. Illustrated........ ....net 2 25

BOWIE (AUG. J., Jun., M. E.). A Practical Treatise on Hydraulic Mining in California. With Description of the Use and Construction of Ditches, Flumes, Wrought-iron Pipes and Dams; Flow of Water on Heavy Grades, and its Applicability, under High Pressure, to Mining. Fifth edition. Small quarto, cloth. Illustrated............................ ...... 5 00

BURGH (N. P.). Modern Marine Engineering, applied to Paddle and Screw Propulsion. Consisting of 36 colored plates, 259 practical woodcut illustrations, and 403 pages of descriptive matter. The whole being an exposition of the present practice of James Watt & Co., J. & G. Rennie, R. Napier & Sons, and other celebrated firms. Thick quarto, half morocco. 10 00

BURT (W. A.). Key to the Solar Compass, and Surveyor's Companion. Comprising all the rules necessary for use in the field; also description of the Linear Surveys and Public Land System of the United States. Notes on the Barometer, Suggestions for an Outfit for a Survey of Four Months, etc. Seventh edition. Pocket-book form, tuck............ 2 50

CALDWELL, (G. C.), and A. A. BRENEMAN. Manual of Introductory Chemical Practice. For the use of Students in Colleges and Normal and High Schools. Fourth edition, revised and corrected. 8vo, cloth Illustrated.................. .... 1 50

CAMPIN (FRANCIS). On the Construction of Iron Roofs. A Theoretical and Practical Treatise, with wood cuts and Plates of Roofs recently executed. 8vo, cloth................................. .... 2 00

CARTER (E. T.). Motive Power and Gearing for Electrical Machinery. A Treatise on the Theory and Practice of the Mechanical Equipment of Power Stations for Electric supply and for Electic Traction. 8vo, cloth. Illustrated........... .... 5 00

CATHCART (Prof. WM. L.). Machine Elements: Shrinkage and Pressure Joints. With tables and diagrams. 8vo, cloth. Illustrated............. ....net 2 50
—— Marine Engine Design........ ....(In Press.)

CHAMBER'S MATHEMATICAL TABLES, consisting of logarithms of Numbers 1 to 108,000, Trigonometrical, Nautical, and other tables. New edition. 8vo, cloth..................... .... 1 75

CHAUVENET (Prof. W.). New Method of Correcting Lunar Distances, and Improved Method of Finding the Error and Rate of a Chronometer, by Equal Altitudes. 8vo, cloth........ .... 2 00

CHRISTIE (W. WALLACE). Chimney Design and Theory. A Book for Engineers and Architects, with numerous half-tone illustrations and plates of famous chimneys. 12mo, cloth..................... .... 3 00

CHURCH (JOHN A.). Notes of a Metallurgical Journey in Europe. 8vo, cloth... .... 2 00

CLARK D. (KINNEAR, C. E.). A Manual of Rules, Tables and Data for Mechanical Engineers. Based

on the most recent investigations. Illustrated with numerous diagrams. 1,012 pages. 8vo, cloth. Sixth edition 5 00
Half morocco... 7 50

CLARK D. (KINNEAR, C. E.). Fuel; its Combustion and Economy, consisting of abridgements of Treatise on the Combustion of Coal. By C. W. Williams; and the Economy of Fuel, by T. S. Prideaux. With extensive additions in recent practice in the Combustion and Economy of Fuel, Coal, Coke, Wood, Peat, Petroleum, etc. Fourth edition. 12mo, cloth.... .... 1 50

—— The Mechanical Engineer's Pocket-book of Tables, Formulæ, Rules and Data. A Handy Book of Reference for Daily Use in Engineering Practice. 16mo, morocco. Second edition................ 3 00

——Tramways, their Construction and Working, embracing a comprehensive history of the system, with accounts of the various modes of traction, a description of the varieties of rolling stock, and ample details of Cost and Working Expenses. Second edition. Re-written and greatly enlarged, with upwards of 400 illustrations. Thick 8vo. cloth. 9 00

—— The Steam Engine. A Treatise on Steam Engines and Boilers; comprising the Principles and Practice of the Combustion of Fuel, the Economical Generation of Steam, the Construction of Steam Boilers, and the Principles, Construction and Performance of Steam Engines, Stationary, Portable, Locomotive and Marine, exemplified in Engines and Boilers of recent date. 1,300 figures in the text, and a series of folding plates drawn to scale. 2 vols. 8vo, cloth. 15 00

CLARK (JACOB M.). A new System of Laying Out Railway Turn-outs instantly, by inspection from Tables. 12mo, leatherette.... .... 1 00

CLAUSEN-THUF (W.). The A. B. C. Universal Commercial Electric Telegraphic Code; especially adapted for the use of Financiers, Merchants, Shipowners, Brokers, Agent, etc. Fourth edition. 8vo, cloth. 5 00
Fifth edition of same 7 00

—— The A1 Universal Commercial Electric Telegraphic Code. Over 1,240 pp., and nearly 90,000 variations. 8vo, cloth.. 7 50

CLEEMANN (THOS. M.). The Railroad Engineer's Practice. Being a Short but Complete Description of the Duties of the Young Engineer in the Preliminary and Location Surveys and in Construction. Fourth edition. Revised and enlarged. Illustrated. 12mo, cloth.................................... . 1 50

CLEVENGER (S. R.). A Treatise on the Method of Government Surveying as prescribed by the U. S. Congress and Commissioner of the General Land Office, with complete Mathematical, Astronomical and Practical Instructions for the use of the United States Surveyors in the field. 16mo, morocco......... 2 50

COFFIN (Prof. J. H. C.). Navigation and Nautical Astronomy. Prepared for the use of the U. S. Naval Academy. New Edition. Revised by Commander Charles Belknap. 52 woodcut illustrations. 12mo, cloth .... net. 3 50

COLE (R. S., M. A.). A Treatise on Photographic Optics. Being an account of the Principles of Optics, so far as they apply to Photography. 12mo, cloth, 103 illustrations and folding plates........... 2 50

COLLINS (JAS. E.). The private Book of Useful Alloys, and Memoranda for Goldsmiths, Jewelers, etc. 18mo, cloth 50

CORNWALL (Prof. H. B.). Manual of Blow-pipe Analysis, Qualitative and Quantitative. With a Complete System of Determinative Mineralogy. 8vo, cloth. With many illustrations...................... 2 50

CRAIG (B. F.). Weights and Measures. An Account of the Decimal System, with Tables of Conversion for Commercial and Scientific Uses. Square 32mo, limp cloth.. 50

CROCKER (F. B.). Electric Lighting. A Practical Exposition of the Art, for use of Engineers, Students, and others interested in the Installation or Operation of Electrical Plants. Sixth edition, revised. 8vo, cloth. Vol. I. The Generating Plant 3 00
Vol. II. Distributing Systems and Lamps. Fourth edition.. 3 00

CROCKER, (F. B.), and S. S. WHEELER. The Practical Management of Dynamos and Motors. Fourth edition (twelfth thousand). Revised and enlarged. With a special chapter by H. A. Foster. 12mo, cloth. Illustrated..................... 1 00

CUMMING (LINNÆUS, M. A.). Electricity treated Experimentally. For the use of Schools and Students New edition. 12mo, cloth........................ 1 50

DAVIES (E. H.). Machinery for Metalliferous Mines. A Practical Treatise for Mining Engineers, Metallurgists and Managers of Mines. With upwards of 400 illustrations. Second edition, rewritten and enlarged. 8vo, cloth..........................net 8 00

DAY (CHARLES). The Indicator and its Diagrams. With Chapters on Engine and Boiler Testing; Including a Table of Piston Constants compiled by W. H. Fowler. 12mo, cloth. 125 illustrations...... 2 00

DERR (W. L.). Block Signal Operation. A Practical Manual. Oblong, cloth........................... 1 50

DIXON (D. B.). The Machinist's and Steam Engineer's Practical Calculator. A Compilation of Useful Rules and Problems arithmetically solved, together with General Information applicable to Shop-Tools, Mill-Gearing, Pulleys and Shafts, Steam-Boilers and Engines. Embracing valuable Tables and Instruction in Screw-cutting, Valve and Link Motion, etc. 16mo, full morocco, pocket form.................. 1 25

DODD (GEO.). Dictionary of Manufactures, Mining, Machinery, and the Industrial Arts. 12mo, cloth ... 1 50

DORR (B. F.). The Surveyor's Guide and Pocket Table Book. 18mo, morocco flaps. Fifth edition, revised, with a second appendix................... 2 00

DRAPER (C. H.). An Elementary Text Book of Light, Heat and Sound, with Numerous Examples. Fourth edition. 12mo, cloth. Illustrated.......... 1 00

—— Heat and the Principles of Thermo-Dynamics. With many illustrations and numerical examples. 12mo, cloth.. 1 50

DUBOIS (A. J.). The New Method of Graphic Statics. With 60 illustrations. 8vo, cloth.......... 1 50

EDDY (Prof. H. T.). Researches in Graphical Statics. Embracing New Constructions in Graphical Statics, a New General Method in Graphical Statics, and the Theory of Internal Stress in Graphical Statics. 8vo, cloth........................ .. 1 50

—— Maximum Stresses under Concentrated Loads. Treated graphically. Illustrated. 8vo, cloth....... 1 50

EISSLER (M.). The Metallurgy of Gold; a Practical Treatise on the Metallurgical Treatment of Gold-Bearing Ores, including the Processes of Concentration and Chlorination, and the Assaying, Melting and Refining of Gold. Fifth Edition, revised and greatly enlarged. 187 illustrations. 12mo, cloth.... 7 50

——The Metallurgy of Silver; a Practical Treatise on the Amalgamation, Roasting and Lixivation of Silver Ores, including the Assaying, Melting and Refining of Silver Bullion. 124 illustrations. Second edition, enlarged, 12mo, cloth 4 00

——The Metallurgy of Argentiferous Lead; a Practical Treatise on the Smelting of Silver Lead Ores and the Refining of Lead Bullion. Including Reports on Various Smelting Establishments and Descriptions of Modern Smelting Furnaces and Plants in Europe and America. With 183 illustrations. 8vo, cloth 5 00

—— Cyanide Process for the Extraction of Gold and its Practical Application on the Witwatersrand Gold Fields in South Africa. Third edition, revised and enlarged. 8vo, cloth. Illustrations and folding plates 3 00

—— A Hand-book on Modern Explosives, being a Practical Treatise on the Manufacture and use of Dynamite, Gun Cotton, Nitro-Glycerine and other Explosive Compounds, including the manufacture of Collodion-Cotton, with chapters on explosives in practical application. Second edition, enlarged with 150 illustrations. 12mo, cloth.......... .... 5 00

ELIOT (C. W.), and STORER (F. H.). A compendious Manual of Qualitative Chemical Analysis. Revised with the co-operation of the authors, by Prof. William R. Nichols. Illustrated. Twentieth edition, newly revised by Prof. W. B. Lindsay. 12mo, cloth......................................net 1 25

ELLIOT (Maj. GEO. H.). European Light-House Systems. Being a Report of a Tour of Inspection made in 1873. 51 engravings and 21 woodcuts. 8vo, cloth 5 00

ELLISON, (LEWIS M.). Practical Application of the Indicator. With reference to the adjustment of Valve Gear on all styles of Engines. Second edition, revised. 8vo. cloth, 100 illustrations............... 2 00

EVERETT (J. D.). Elementary Text-book of Physics. Illustrated. Seventh edition. 12mo, cloth . 50

EWING (Prof. A. J.). The Magnetic Induction in Iron and other metals. 159 illustrations. 8vo, cloth 4 00

FANNING (J. T.). A Practical Treatise on Hydraulic and Water-Supply Engineering. Relating to the Hydrology, Hydro-dynamics, and Practical Construction of Water-Works in North America. 180 illustrations. 8vo, cloth. Fifteenth edition, revised, enlarged, and new tables and illustrations added. 650 pages............................... 5 00

FISH (J. C. L.). Lettering of Working Drawings. Thirteen plates, with descriptive text. Oblong, 9x12½, boards..................................... 1 00

FISKE (Lieut. BRADLEY A., U. S. N.). Electricity in Theory and Practices or, The Elements of Electrical Engineering. Eighth edition. 8vo, cloth 2 50

FISHER (H. K. C. and DARBY, W. C.). Students' Guide to Submarine Cable Testing. 8vo, cloth...... 2 50

FISHER (W. C.). The Potentiometer and its Adjuncts. 8vo, cloth....................................... 2 25

FLEISCHMANN (W.). The Book of the Dairy. A Manual of the Science and Practice of Dairy Work. Translated from the German, by C. M. Aikman and R. Patrick Wright. 8vo, cloth..................... 4 00

FLEMING (Prof. J. A.). The Alternate Current Transformer in Theory and Practice. Vol. 1—The Induction of Electric Currents; 611 pages. New edition. Illustrated. 8vo, cloth................... 5 00
Vol. 2. The Utilization of Induced Currents. Illustrated. 8vo, cloth 5 00

—— - Electric Lamps and Electric Lighting. Being a course of four lectures delivered at the Royal Institution, April–May, 1894. 8vo, cloth, fully illustrated 3 00

FLEMING (Prof. J. A.). Electrical Laboratory Notes and Forms, Elementary and advanced. 4to, cloth, illustrated..................................... 5 00

FOLEY (NELSON), and THOS. PRAY, Jr. The Mechanical Engineers' Reference Book for Machine and Boiler Construction, in 2 parts. Part 1—General Engineering Data. Part 2—Boiler Construction. With fifty-one plates and numerous illustrations, specially drawn for this work. Folio, half mor.....25 00

FORNEY (MATTHIAS N.). Catechism of the Locomotive. Second edition, revised and enlarged. Forty-sixth thousand. 8vo, cloth.................. 3 50

FOSTER (Gen. J. G., U. S. A.). Submarine Blasting in Boston Harbor, Massachusetts. Removal of Tower and Corwin Rocks. Illustrated with 7 plates 4to, cloth.. 3 50

FOSTER (H. A.). Electrical Engineers' Pocket Book. 1000 pages with the collaboration of Eminent Specialists. Third edition, revised. Pocket size, full leather... 5 00

FOSTER (JAMES). Treatise on the Evaporation on Saccharine, Chemical and other Liquids by the Multiple System in Vacuum and Open Air. Second edition. Diagrams and large plates. 8vo, cloth.... 7 50

FOWLER. Mechanical Engineers' Pocket Book for 1905.. 1 00

FOX (WM.), and C. W. THOMAS, M. E. A Practical Course in Mechanical Drawing, second edition, revised. 12mo, cloth, with plates.......... 1 25

FRANCIS (JAS. B., C. E.). Lowell Hydraulic Experiments. Being a selection from experiments on Hydraulic Motors, on the Flow of Water over Weirs, in open Canals of uniform rectangular section, and through submerged Orifices and diverging Tubes. Made at Lowell, Mass. Fourth edition, revised and enlarged, with many new experiments, and illustrated with 23 copper-plate engravings. 4to, cloth...................................15 00

FROST (GEO. H.). Engineer's Field Book. By C. S. Cross. To which are added seven chapters on Railroad Location and Construction. Fourth edition. 12mo, cloth.. 1 00

FULLER (GEORGE W.). Report on the Investigations into the Purification of the Ohio River Water at Louisville, Kentucky, made to the President and Directors of the Louisville Water Company. Published under agreement with the Directors. 4to, cloth. 8 full page plates........................net 10 00

GEIPEL (WM.), and KILGOUR, (M. H.) A Pocketbook of Electrical Engineering Formula. Illustrated. 18mo, morocco........................ 3 00

GERBER (NICHOLAS). Chemical and Physical Analysis of Milk, Condensed Milk and Infant's Milk-Food. 8vo, cloth. 1 25

GESCHWIND (LUCIEN). Manufacture of Alum and Sulphates, and other Salts of Alumina and Iron; their uses and applications as mordants in dyeing and calico printing, and their other applications in the Arts, Manufactures, Sanitary Engineering, Agriculture, and Horticulture. Translated from the French by Charles Salter. With tables, figures and diagrams. 8vo, cloth, illus.................net 5 00

GIBBS (WILLIAM E.). Lighting by Acetylene, Generators, Burners and Electric Furnaces. With 66 illustrations. Second edition revised. 12mo, cloth.. 1 50

GILLMORE (GEN. Q. A.). Treatise on Limes, Hydraulic Cements, and Mortars. Papers on Practical Engineering, United States Engineer Department, No. 9, containing Reports of numerous Experiments conducted in New York City during the years of 1858 to 1861, inclusive. With numerous illustrations. 8vo cloth.... .... 4 00

—— Practical Treatise on the Construction of Roads. Streets, and Pavements, with 70 illustrations. 12mo, cloth.................................. 2 00

—— Report on Strength of Building Stones in the United States, etc. 8vo, illustrated cloth.......... 1 00

GOLDING (HENRY A.). The Theta-Phi Diagram. Practically applied to Steam, Gas, Oil and Air Engines. 12mo, cloth. Illustrated................net 1 25

GOODEVE (T. M.). A Text-Book on the Steam Engine. With a Supplement on Gas-Engines. Twelfth Edition, enlarged. 143 illustrations. 12mo, cloth... 2 00

GORE (G., F. R. S.). The Art of Electrolytic Separation of Metals, etc. (Theoretical and Practical.) Illustrated. 8vo, cloth........................ 3 50

GOULD (E. SHERMAN). The Arithmetic of the
Steam Engine. 8vo, cloth............................. 1 00
GRIFFITHS (A. D., Ph. D.). A Treatise on Manures,
or the Philosophy of Manuring. A Practical Hand-
Book for the Agriculturist, Manufacturer and
Student. 12mo, cloth.................................. 3 00
GROVER (FREDERICK). Practical Treatise on
Modern Gas and Oil Engines. 8vo, cloth. Illustrated 2 00
GURDEN (RICHARD LLOYD). Traverse Tables:
computed to 4 places Decimals for every ° of angle
up to 100 of Distance. For the use of Surveyors and
Engineers. New Edition. Folio, half morocco. .. 7 50
GUY ARTHUR (F.). Electric Light and Power,
giving the Result of Practical Experience in Central
Station Work. 8vo, cloth. Illustrated............. 2 50
HAEDER (HERMAN C. E.). A Hand-book on the
Steam Engine. With especial reference to small
and medium sized engines. English edition re-edited
by the author from the second German edition, and
translated with considerable additions and alterations by H. H. P. Powels. 12mo, cloth. Nearly 1100
illustrations... 3 00
HALL (WM. S. Prof.). Elements of the Differential
and Integral Calculus. Second edition. 8vo, cloth.
Illustrated.....................................net 2 25
HALSEY (F. A.). Slide Valve Gears; an Explanation of
the action and Construction of Plain and Cut-off
Slide Valves. Illustrated. 12mo, cloth. Sixth
edition... 1 50
—— The Use of the Slide Rule. Illustrated with
diagrams and folding plates. 16mo, boards........ 50
HAMILTON (W. G.). Useful Information for Railway Men. Tenth Edition, revised and enlarged.
562 pages, pocket form. Morocco, gilt............. 2 00
HANCOCK (HERBERT). Text Book of Mechan-
ics and Hydrostatics, with over 500 diagrams. 8vo,
cloth... 1 75
HARRISON (W. B.). The Mechanics' Tool Book.
With Practical Rules and Suggestions for use of
Machinists, Iron-Workers, and others. Illustrated
with 44 engravings. 12mo, cloth.................... 1 50

HASKINS (C. H.). The Galvanometer and its Uses. A Manual for Electricians and Students. Fourth edition. 12mo, cloth.................................. 1 50

HAWKE (WILLIAM H.). The Premier Cipher Telegraphic Code Containing 100,000 Words and Phrases. The most complete and most useful general code yet published. 4to, cloth...................... 5 00

—— 100,000 Words; Supplement to the Premier Code. All the words are selected from the official vocabulary. Oblong quarto, cloth.............................. 4 20

HAWKINS (C. C.) and WALLIS (F.). The Dynamo; its Theory, Design and Manufacture. 190 illustrations, 12mo, cloth... 3 00

HAY (ALFRED). Principles of Alternate-Current Working. 12mo, cloth, illustrated...................... 2 00

HEAP (Major D. P., U. S. A.). Electrical Appliances of the Present Day. Report of the Paris Electrical Exposition of 1881. 250 illustrations. 8vo, cloth..... 2 00

HEAVISIDE (OLIVER). Electromagnetic Theory. 8vo, cloth, two volumes each........................ 5 00

HENRICI (OLAUS). Skeleton Structures, Applied to the Building of Steel and Iron Bridges. Illustrated 1 50

HERRMANN (GUSTAV). The Graphical Statics of Mechanism. A Guide for the Use of Machinists, Architects, and Engineers; and also a Text-book for Technical Schools. Translated and annotated by A. P. Smith, M. E. 12mo, cloth, 7 folding plates. Third Edition... 2 00

HERMANN (FELIX). Painting on Glass and Porcelain and Enamel Painting. On the Basis of Personal Practical Experience of the Condition of the Art up to date. Translated by Charles Salter. Second greatly enlarged edition. 8vo, cloth, Illustrations, net... 3 50

HEWSON (WM.). Principles and Practice of Embanking Lands from River Floods, as applied to the Levees of the Mississippi. 8vo, cloth.............. 2 00

HILL (JOHN W.). The Purification of Public Water Supplies. Illustrated with valuable Tables, Diagrams and Cuts. 8vo, cloth, 304 pages.............. 3 00

—— The Interpretation of Water Analyses..(In Press)

SCIENTIFIC PUBLICATIONS. 17

HOBBS (W. R. P.). The Arithmetic of Electrical Measurements with numerous examples. Fully Worked, 12mo, cloth............................... .50

HOFF (WM. B., Com. U. S. Navy.). The Avoidance of Collisions at Sea. 18mo, morocco .... .75

HOLLEY (ALEXANDER L.). Railway Practice. American and European Railway practice in the Economical Generation of Steam. 77 lithographed plates. Folio, cloth 12 00

HOLMES (A. BROMLEY). The Electric Light Popularly Explained. Fifth Edition. Illustrated. 12mo, paper................................. 50

HOPKINS (NEVIL M.). Model Engines and small Boats. New Methods of Engine and Boiler Making with a chapter on Elementary Ship Design and Construction. 12mo, cloth......... .... 1 25

HOSPITALIER (E.). Polyphased Alternating Currents. Illustrated. 8vo, cloth. 1 40

HOWARD (C. R.). Earthwork Mensuration on the Basis of the Prismoidal Formulae. Containing Simple and Labor-saving Methods of obtaining Prismoidal Contents directly from End Areas. Illustrated by Examples and accompained by Plain Rules for Practical Uses. Illustrated. 8vo, cloth... ..... 1 50

HUMBER (WILLIAM, C. E.). A Handy Book for the Calculation of Strains in Girders and Similar Structures, and their Strength; Consisting of Formulae and Corresponding Diagrams, with numerous details for practical application, etc. Fourth Edition. 12mo, cloth...... .... 2 50

HURST (GEORGE H.). Colour; A Hand-book of the Theory of Colour. Containing ten coloured plates and 72 diagrams. 8vo, cloth. Illustrated. Price.... 2 50

—— Lubricating Oils, Fats and Greases. Their Origin, Preparation, Properties, Uses and Analysis. 313 pages, with 65 illustrations. 8vo, cloth............ 3 00

—— Soaps; A Practical Manual of the Manufacture of Domestic, Toilet and other Soaps. Illustrated with 66 Engravings. 8vo, cloth........................ 5 00

HUTCHINSON (W. B.). Patents and How to Make Money out of Them. Member of New York Bar. 12mo, cloth. New York, 1899 .... 1 25

HUTTON (W. S.). Steam Boiler Construction. A Practical Hand-book for Engineers, Boiler Makers and Steam Users. Containing a large collection of rules and data relating to recent practice in the design, construction, and working of all kinds of stationary, locomotive and marine steam boilers. With upwards of 500 illustrations. Third edition. Carefully revised and much enlarged. 8vo, cloth... 6 00

—— Practica Engineer's Hand-book, Comprising a treatise on Modern Engines and Boilers, Marine, Locomotive and Stationary. Fourth edition. Carefully revised with additions. With upwards of 570 illustrations. 8vo, cloth........................... 7 00

—— The Works' Manager's Hand-book of Modern Rules, Tables, and Data for Civil and Mechanical Engineers. Millwrights and Boiler Makers, etc., etc. With upwards of 150 illustrations. Fifth edition. Carefully revised, with additions. 8vo, cloth....... 6 00

INNES (CHARLES H.). Problems in Machine Design. For the Use of Students, Draughtsmen and others. 12mo, cloth.. 1 50

—— Centrifugal Pumps, Turbines and Water Motors. Including the Theory and Practice of Hydraulics. 12mo, cloth.................................... net 2 00

ISHERWOOD (B. F.). Engineering Precedents for Steam Machinery. Arranged in the most practical and useful manner for Engineers. With illustrations. 2 vols. in 1. 8vo, cloth.................. .... 2 50

JAMESON (CHARLES D.). Portland Cement. Its Manufacture and Use. 8vo, cloth................. ...... 1 50

JAMIESON (ANDREW C. E.). A Text-Book on Steam and Steam Engines. Specially arranged for the use of Science and Art, City and Guilds of London Institute, and other Engineering Students. Tenth edition. Illustrated. 12mo, cloth................. 3 00

—— Elementary Manual on Steam and the Steam Engine. Specially arranged for the use of First-Year Science and Art, City and Guilds of London Institute, and other Elementary Engineering Students. Third edition. 12mo, cloth 1 50

JANNETTAZ (EDWARD). A Guide to the Determination of Rocks: being an Introduction to Lithology. Translated from the French by G. W. Plympton, Professor of Physical Science at Brooklyn Polytechnic Institute. 12mo, cloth............ 1 50

JOHNSTON, Prof. J. F. W., and CAMERON, Sir CHAS. Elements of Agricultural Chemistry and Geology. Seventeenth edition. 12mo, cloth........ 2 60

JOYNSON (F. H.). The Metals used in Construction. Iron, Steel, Bessemer Metal, etc. Illustrated. 12mo, cloth 75

—— Designing and Construction of Machine Gearing. Illustrated. 8vo, cloth.. .... 2 00

KANSAS CITY BRIDGE (THE.) With an Account of the Regimen of the Missouri River and a Description of the Methods used for Founding in that River. By O. Chanute, Chief Engineer, and George Morrison, Assistant Engineer. Illustrated with 5 lithographic views and 12 plates of plans. 4to, cloth.... 6 00

KAPP (GISBERT C. E.). Electric Transmission of Energy and its Transformation, Subdivision, and Distribution. A Practical Hand-book. Fourth edition, revised. 12mo, cloth.................·......... 3 50

—— Dynamos, Alternators and Transformers. 138 Illustrations. 12mo, cloth......... .... 4 00

KEMPE (H. R.). The Electrical Engineer's Pocket-Book of Modern Rules, Formulæ, Tables and Data. Illustrated. 32mo, mor. gilt...................... 1 75

KENNELLY (A. E.). Theoretical Elements of Electro-Dynamic Machinery. 8vo, cloth...... .... 1 50

KILGOUR, M. H., SWAN, H., and BIGGS, C. H. W. Electrical Distribution; its Theory and Practice. 174 Illustrations. 12mo, cloth..................... 4 00

KING (W. H.). Lessons and Practical Notes on Steam. The Steam Engine, Propellers, etc., for Young Marine Engineers, Students, and others. Revised by Chief Engineer J. W. King, United States Navy. Nineteenth edition, enlarged. 8vo, cloth.... 2 00

KINGDON (J. A.). Applied Magnetism. An introduction to the Design of Electromagnetic Apparatus. 8vo. cloth....................... .... 3 00

KIRKALDY (WM. G.). Illustrations of David Kirkaldy's System of Mechanical Testing, as Originated and Carried On by him during a Quarter of a Century. Comprising a Large Selection of Tabulated Results, showing the Strength and other Properties of Materials used in Construction, with explanatory Text and Historical Sketch. Numerous engravings and 25 lithographed plates. 4to, cloth...................20 00

KIRKWOOD (JAS. P.). Report on the Filtration of River Waters for the supply of Cities, as practised in Europe, made to the Board of Water Commissioners of the City of St. Louis. Illustrated by 30 double-plate engravings. 4to, cloth...................... 7 50

KNIGHT (AUSTIN M., *Lieutenant-Commander, U. S. N.*). Modern Seamanship. Illustrated with 136 full-page plates and diagrams. 8vo, cloth. Second edition, revised...net 6 00
Half morocco.. .. 7 50

LARRABEE (C. S.). Cipher and Secret Letter and Telegraphic Code, with Hog's Improvements. The most perfect Secret Code ever invented or discovered. Impossible to read without the key. 18mo, cloth 60

LEASK (A. RITCHIE). Breakdowns at Sea and How to Repair Them. With eighty-nine Illustrations. 8vo, cloth. Second edition.................. 2 00

—— Triple and Quadruple Expansion Engines and Boilers and their Management. With fifty-nine illustrations. Third edition, revised. 12mo, cloth.. 2 00

—— Refrigerating Machinery: Its Principles and Management. With sixty-four illustrations. 12mo, cloth 2 00

LECKY (S. T. S.). "Wrinkles" in Practical Navigation. With 130 illustrations. 8vo, cloth. Ninth edition, revised.....................................net 8 00

LEVY (C. L.). Electric Light Primer. A Simple and Comprehensive Digest of all of the most important facts connected with the running of the dynamo, and electric lights, with precautions for safety. For the use of persons whose duty it is to look after the plant. 8vo, paper.................................... 50

LIVACHE (ACH., Ingenieur Civil Des Mines).
The Manufacture of Varnishes, Oil Crushing, Refining and Boiling and Kindred Industries. Translated from the French and greatly extended, by John Geddes McIntosh. 8vo, cloth. Illustrated.......net 5 00

LOCKE (ALFRED G., and CHARLES G.) A Practical Treatise on the Manufacture of Sulphuric Acid. With 77 Constructive Plates drawn to Scale Measurements, and other Illustrations. Royal 8vo, cloth.10 00

LOCKERT (LOUIS). Petroleum Motor-Cars. 12mo, cloth................................. 1 50

LOCKWOOD (THOS. D.). Electricity, Magnetism, and Electro-Telegraphy. A Practical Guide for Students, Operators, and Inspectors. 8vo, cloth. Third edition................................... 2 50

—— Electrical Measurement and the Galvanometer; its Construction and Uses. Second edition. 32 illustrations. 12mo, cloth............................. 1 50

LODGE (OLIVER J.). Elementary Mechanics, including Hydrostatics and Pneumatics. Revised edition. 12mo, cloth............................. 1 50

LORING (A. E.). A Hand book of the Electro-Magnetic Telegraph. Fourth edition, revised............

LUCE (Com. S. B.). Text-Book of Seamanship. The Equipping and Handling of Vessels under Sail or Steam. For the use of the U. S. Naval Academy. Revised and enlarged edition, by Lt. Wm. S. Benson. 8vo, cloth..10 00

LUNGE (GEO.). A Theoretical and Practical Treatise on the Manufacture of Sulphuric Acid and Alkali with the Collateral Branches. Vol. I. Sulphuric Acid, Second edition, revised and enlarged. 342 Illustrations. 8vo, cloth........................15 00
Vol. II. Second edition, revised and enlarged. 8vo, cloth....................................16 80
Vol III. 8vo, cloth. New edition, 1896...........15 00

LUNGE. (GEO.), and HURTER, F. The Alkali Maker's Pocket-Book. Tables and Analytical Methods for Manufacturers of Sulphuric Acid, Nitric Acid, Soda, Potash and Ammonia. Second edition. 12mo, cloth 3 00

LUQUER (LEA McILVAINE, Ph. D.). Minerals in Rock Sections. The Practical Method of Identifying Minerals in Rock Sections with the microscope, Especially arranged for Students in Technical and Scientific Schools. 8vo. cloth. Illustrated..net 1 50

MACKROW (CLEMENT). The Naval Architect's and Ship-Builder's Pocket-Book of Formulæ, Rules and Tables; and Engineers' and Surveyors' Handy-Book of Reference. Eighth edition, revised and enlarged. 16mo, limp leather. Illustrated. 5 00

MAGUIRE (Capt. EDWARD. U. S. A.). The Attack and Defence of Coast Fortifications. With Maps and Numerous Illustrations. 8vo, cloth...... 2 50

MAGUIRE (WM. R.). Domestic Sanitary Drainage and Plumbing Lectures on Practical Sanitation. 332 illustrations. 8vo.......................... 4 00

MARKS (EDWARD C. R.). Mechanical Engineering Materials: their Properties and Treatment in Construction. 12mo, cloth. Illustrated................ 60

—— Notes on the Construction of Cranes and Lifting Machinery. 12mo, cloth........................... 1 50

MARKS (G. C.). Hydraulic Power Engineering: a Practical Manual on the Concentration and Transmission of Power by Hydraulic Machinery. With over 200 diagrams, figures, and tables. 8vo, cloth. Illustrated................................. 3 50

MAVER (WM.). American Telegraphy: Systems, Apparatus, Operation. 450 illustrations. 8vo, cloth. 3 50

MAYER (Prof. A. M.). Lecture Notes on Physics. 8vo, cloth.. 2 00

McCULLOCH (Prof. R. S.). Elementary Treatise on the Mechanical Theory of Heat, and its application to Air and Steam Engines. 8vo, cloth.......... 3 50

McNEILL (BEDFORD). McNeill's Code. Arranged to meet the requirements of Mining, Metallurgical and Civil Engineers, Directors of Mining, Smelting and other Companies, Bankers, Stock and Share Brokers, Solicitors, Accountants, Financiers, and General Merchants. Safety and Secrecy. 8vo, cloth. 6 00

McPHERSON (J. A., *A. M. Inst. C. E.*). Waterworks Distribution: a practical guide to the laying out of systems of distributing mains for the supply of water to cities and towns. With tables, folding plates and numerous full-page diagrams. 8vo, cloth. Illustrated.................................... 2 50

METAL TURNING. By a Foreman Pattern Maker. Illustrated with 81 engravings. 12mo, cloth........ 1 50

MINIFIE (WM.). Mechanical Drawing. A Textbook f Geometrical Draw ng for the use of Mechanics and Schools, in which the Definitions and Rules of Geometry are familiarly explained; the Practical Problems are arranged from the most simple to the more complex, and in their description technicalities are avoided as much as possible. With illustrations for Draw ng Plans, Sections, and Elevations of Railways and Machinery; an Introduction to Isometrical Drawing, and an Essay on Linear Perspective and Shadows Illustrated with over 200 diagrams engraved on steel. Ninth thousand. With an appendix on the Theory and Application of Colors. 8vo, cloth.. 4 00

—— Geometrical Drawing. Abridged from the Octavo edition, for the use of schools. Illustrated with 48 steel plates. Ninth edition. 12mo, cloth........... 2 00

MODERN METEOROLOGY. A Series of Six Lectures, delivered under the auspices of the Meteorological Society in 1870. Illustrated. 12mo, cloth . 1 50

MOREING (C. A.), and NEAL (THOMAS). Telegraphic Mining Code Alphabetically arranged. Second edition. 8vo, cloth....................... 8 40

MORRIS (E.). Easy Rules for the Measurement of Earthworks by means of the Prismoidal Formula. 8vo, cloth. Illustrated............................. 1 50

MOSES (ALFRED J.), and PARSONS, C. L. Elements of Mineralogy, Crystallography and Blowpipe Analysis from a practical standpoint. Fourth thousand. 8vo, cloth. 366 illustrations...... ...net 2 50

MOSES (ALFRED J.). The Characters of Crystals. An Introduction to Physical Crystallography, containing 321 Illustrations and Diagrams. 8vo, 211 pp..net 2 00

MOORE (E. C. S.). New Tables for the Complete Solution of Ganguillet and Kutter's Formula for the flow of liquids in open channels, pipes, sewers and conduits. In two parts. Part I, arranged for 1,080 inclinations from 1 over 1 to 1 over 21,120 for fifteen different values of (*n*). Part II, for use with all other values of (*n*). With large folding diagram. 8vo, cloth. Illustrated..........................net 5 00

MULLIN (JOSEPH P., M. E.). Modern Moulding and Pattern-Making. A Practical Treatise upon Pattern-Shop and Foundry Work; embracing the Moulding of Pulleys, Spur Gears, Worm Gears, Balance-Wheels, Stationary Engine and Locomotive Cylinders, Globe Valves, Tool Work, Mining Machinery, Screw Propellers, Pattern-Shop Machinery, and the latest improvements in English and American Cupolas; together with a large collection of original and carefully selected Rules and Tables for everyday use in the Drawing Office, Pattern-Shop and Foundry. 12mo, cloth. Illustrated............... 2 50

MUNRO (JOHN C. E.), and JAMIESON ANDREW. C. E. A Pocket-book of Electrical Rules and Tables for the use of Electricians and Engineers. Thirteenth edition, revised and enlarged. With numerous diagrams. Pocket size. Leather.. 2 50

MURPHY (J. G., M. E.). Practical Mining. A Field Manual for Mining Engineers. With Hints for Investors in Mining Properties. 16mo, morocco tucks.. 1 00

NAQUET (A.). Legal Chemistry. A Guide to the Detection of Poisons, Falsification of Writings, Adulteration of Alimentary and Pharmaceutical Substances, Analysis of Ashes, and examination of Hair, Coins, Arms, and Stains, as applied to Chemical Jurisprudence. Translated from the French, by J. P. Battershall, Ph. D., with a preface by C. F. Chandler, Ph. D., M. D., LL. D. 12mo, cloth........ 2 00

NASMITH (JOSEPH). The Student's Cotton Spinning. Third edition, revised and enlarged. 8vo, cloth. 622 pages. 250 Illustrations................ 3 00

NEWALL (JOHN W.). Plain Practical Directions for Drawing, Sizing and Cutting Bevel-Gears, showing how the Teeth may be cut in a Plain Milling Machine or Gear Cutter so as to give them a correct

SCIENTIFIC PUBLICATIONS. 25

shape from end to end; and showing how to get out all particulars for the Workshop without making any Drawings. Including a Full Set of Tables of Reference. Folding plates. 8vo, cloth............. 1 50

NEUBURGER (HENRY) and HENRI NOALHAT. Technology of Petroleum. The Oil Fields of the World; their History, Geography and Geology. With 153 illustrations and 25 plates. Translated from the French by John Geddes McIntosh. 8vo, cloth................................... net 10 00

NEWLANDS (JAMES). The Carpenters' and Joiners' Assistant: being a Comprehensive Treatise on the Selection, Preparation and Strength of Materials, and the Mechanical Principles of Framing. Illustrated. Folio, half morocco..................15 00

NIPHER (FRANCIS E., A. M.). Theory of Magnetic Measurements, with an appendix on the Method of Least Squares. 12mo, cloth 1 00

NOAD (HENRY M.). The Students' Text-Book of Electricity. A new edition, carefully revised. With an Introduction and additional chapters by W. H. Preece. With 471 illustrations. 12mo, cloth........ 4 00

NUGENT (E.). Treatise on Optics; or, Light and Sight theoretically and practically treated, with the application to Fine Art and Industrial Pursuits. With 103 illustrations. 12mo, cloth. 1 50

O'CONNOR (HENRY). The Gas Engineer's Pocket-Book. Comprising Tables, Notes and Memoranda; relating to the Manufacture, Distribution and Use of Coal Gas and the Construction of Gas Works. 12mo, full leather, gilt edges.................... 3 50

OUDIN (M. A.). Standard Polyphase Apparatus and Systems. With many photo-reproductions, diagrams, and tables. Third edition, revised. 8vo, cloth. Illustrated........................... 3 00

PAGE (DAVID). The Earth's Crust, A Handy Outline of Geology. 16mo, cloth..................... 75

PALAZ (A., ScD.). A Treatise on Industrial Photometry, with special application to Electric Lighting. Authorized translation from the French, by George W. Patterson, Jr. Second edition, revised. 8vo, cloth. Illustrated...................... 4 00

PARSHALL (H. F.) and H. M. HOBART. Armature Windings of Electric Machines. With 140 full-page plates, 65 tables, and 165 pages of descriptive letter-press. 4to, cloth........   .. 7 50

PEIRCE (B.). System of Analytic Mechanics. 4to, cloth................   ....10 00

PERRINE (F. A. C., A.M., D.Sc.). Conductors for Electrical Distribution; their Manufacture and Materials, the Calculation of Circuits, Pole-Line Construction, Underground Working and other Uses. 8vo, cloth. Illustrated........ ....net 3 50
                                                Postage 25

PERRY (JOHN). Applied Mechanics. A Treatise for the use of students who have time to work experimental, numerical and graphical exercises illustrating the subject. 8vo, cloth. 650 pages.,net 2 50

PHILLIPS (JOSHUA). Engineering Chemistry. A Practical Treatise for the use of Analytical Chemists, Engineers, Iron Masters, Iron Founders, students and others. Comprising methods of Analysis and Valuation of the principal materials used in Engineering works, with numerous Analyses, Examples and Suggestions. 314 illustrations. Second edition, revised and enlarged. 8vo, cloth.......... 4 50

PICKWORTH (CHAS. N.). The Indicator Handbook. A Practical Manual for Engineers. Part I. The Indicator: Its Construction and Application. 81 illustrations. 12mo, cloth........................ 1 50

—— The Slide Rule. A Practical Manual of Instruction for all Users of the Modern Type of Slide Rule, exhibiting the Application of the Instrument to the Everyday Work of the Engineer,—Civil, Mechanical and Electrical. 12mo, flexible cloth. Fifth edition. 80

PLANE TABLE (THE). Its Uses in Topographical Surveying. From the Papers of the United States Coast Survey. Illustrated. 8vo, cloth.............. 2 00

PLANTE (GASTON). The Storage of Electrical Energy, and Researches in the Effects created by Currents, combining Quantity with High Tension. Translated from the French by Paul B. Elwell. 89 illustrations. 8vo............. .... 4 00

PLATTNER. Manual of Qualitative and Quantitative Analysis with the Blow-Pipe. Eighth edition, revised. Translated by Henry B. Cornwall, E.M., Ph.D., assisted by John H. Caswell, A.M. From the sixth German edition, by Prof. Friedrich Kolbeck. Illustrated with 87 woodcuts. 463 pages. 8vo, cloth..net 4 00

PLYMPTON (Prof. GEO. W.). The Aneroid Barometer: its Construction and Use. Compiled from several sources. Fourth edition. 16mo, boards. Illustrated... 50

POCKET LOGARITHMS, to Four Places of Decimals, including Logarithms of Numbers, and Logarithmic Sines and Tangents to Single Minutes. To which is added a Table of Natural Sines, Tangents, and Co-Tangents. 16mo, boards..................... 50

POPE (F. L.). Modern Practice of the Electric Telegraph. A Technical Hand-book for Electricians, Managers and Operators. Fifteenth edition, rewritten and enlarged, and fully illustrated. 8vo, cloth. 1 50

POPPLEWELL (W. C.). Elementary Treatise on Heat and Heat Engines. Specially adapted for engineers and students of engineering. 12mo, cloth. Illustrated... 3 00

POWLES (H. H.). Steam Boilers..........(In Press.)

PRAY (Jr., THOMAS). Twenty Years with the Indicator; being a Practical Text-Book for the Engineer or the Student, with no complex Formulæ. Illustrated. 8vo, cloth.............................. 2 50

—— Steam Tables and Engine Constant. Compiled from Regnault, Rankine and Dixon directly, making use of the exact records. 8vo, cloth................ 2 00

PRACTICAL IRON FOUNDING. By the Author of "Pattern Making," &c., &c. Illustrated with over one hundred engravings. 12mo, cloth......... 1 50

PREECE (W. H.). Electric Lamps ....(In Press.)

PREECE (W. H.), and STUBBS, A. T. Manual of Telephony. Illustrations and plates. 12mo, cloth. 4 50

PREMIER CODE. (See Hawk, Wm. H.)

PRESCOTT (Prof. A. B.). Organic Analysis. A Manual of the Descriptive and Analytical Chemistry of certain Carbon Compounds in Common Use; a Guide in the Qualitative and Quantitative Analysis of Organic Materials in Commercial and Pharmaceutical Assays, in the estimation of Impurities under Authorized Standards, and in Forensic Examinations for Poisons, with Directions for Elementary Organic Analysis. Fifth edition. 8vo, cloth........ 5 00

PRESCOTT (Prof. A. B.). Outlines of Proximate Organic Analysis, for the Identification, Separation, and Quantitative Determination of the more commonly occurring Organic Compounds. Fourth edition. 12mo, cloth... .... 1 75

—— First Book in Qualitative Chemistry. Eighth edition. 12mo, cloth 1 50

—— and Otis Coe Johnson. Qualitative Chemical Analysis. A Guide in the Practical Study of Chemistry and in the work of Analysis. Fifth fully revised edition. With Descriptive Chemistry extended throughout.......... ....net 3 50

PRITCHARD (O. G.). The Manufacture of Electric Light Carbons. Illustrated. 8vo, paper 60

PULLEN (W. W. F.). Application of Graphic Methods to the Design of Structures. Specially prepared for the use of Engineers. 12mo, cloth. Illustrated.
net 2 50

PULSIFER (W. H.). Notes for a History of Lead. 8vo, cloth, gilt tops........ ... .·. 4 00

PYNCHON (Prof. T. R.). Introduction to Chemical Physics, designed for the use of Academies, Colleges, and High Schools. Illustrated with numerous engravings, and containing copious experiments with directions for preparing them. New edition, revised and enlarged, and illustrated by 269 illustrations on wood. 8vo, cloth....................... 3 00

RADFORD (Lieut. CYRUS S.). Hand-book on Naval Gunnery. Prepared by Authority of the Navy Department. For the use of U. S. Navy, U. S. Marine Corps and U. S. Naval Reserves. Revised and enlarged, with the assistance of Stokely Morgan, Lieut. U. S. N. Third edition. 12mo, flexible leather. 2 00

RAFTER (GEO. W.), and M. N. BAKER. Sewage Disposal in the United States. Illustrations and folding plates. Second edition. 8vo, cloth........ 6 00

RAM (GILBERT S.). The Incandescent Lamp and its Manufacture. 8vo, cloth...................... 3 00

RANDALL (J. E.). A Practical Treatise on the
Incandescent Lamp. Illustrated. 16mo, cloth..... 50

RANDALL (P. M.). Quartz Operator's Hand-book.
New edition, revised and enlarged, fully illustrated.
12mo, cloth... 2 00

RANKINE (W. J. MACQUORN.) Applied Mechanics. Comprising the Principles of Statics and Cinematics, and Theory of Structures, Mechanism, and Machines. With numerous diagrams. Fifteenth edition. Thoroughly revised by W. J. Millar. 8vo, cloth... 5 00

—— Civil Engineering. Comprising Engineering Surveys, Earthwork, Foundations, Masonry, Carpentry, Metal Work, Roads, Railways, Canals, Rivers, Water Works, Harbors, etc. With numerous tables and illustrations. Twentieth edition. Thoroughly revised by W. J. Millar. 8vo, cloth............... 6 50

—— Machinery and Millwork. Comprising the Geometry, Motions, Work, Strength, Construction, and Objects of Machines, etc. Illustrated with nearly 300 wood cuts. Seventh edition. Thoroughly revised by W. J. Millar. 8vo, cloth....................... 5 00

—— The Steam Engine and other Prime Movers. With diagram of the Mechanical Properties of Steam-folding plates, numerous tables and illustrations. Thirteenth edition. Thoroughly revised by W. J. Millar. 8vo, cloth 5 00

—— Useful Rules and Tables for Engineers and Others. With appendix, tables, tests, and formulæ for the use of Electrical Engineers. Comprising Submarine Electrical Engineering, Electric Lighting, and Transmission of Power. By Andrew Jamieson, C. E., F. R S. E. Seventh edition. Thoroughly revised by W. J. Millar. Crown 8vo, cloth. 4 00

—— A Mechanical Text-Book. By Prof. Macquorn Rankine and E. F. Bamber, C. E. With numerous illustrations. Fourth edition. 8vo, cloth.......... 3 50

RAPHAEL (F. C.). |Localisation of Faults in Electric Light Mains. 8vo, cloth............. .... 2 00

RECKENZAUN (A.). Electric Traction on Railways and Tramways. 213 illustrations. 12mo, cloth. ... 4 00

REED'S ENGINEERS' HAND-BOOK to the Local Marine Board Examinations for Certificates of Competency as First and Second Class Engineers. By W. H. Thorn. With the answers to the Elementary Questions. Illustrated by 297 diagrams and 36 large plates. Seventeenth edition, revised and enlarged. 8vo, cloth.................................... 5 00

—— Key to the Seventeenth Edition of Reed's Engineer's Hand-book to the Board of Trade Examinations for First and Second Class Engineers and containing the working of all the questions given in the examination papers. By W. H. Thorn. 8vo, cloth.. 3 00

—— Useful Hints to Sea-going Engineers, and How to Repair and Avoid "Break Downs;" also Appendices Containing Boiler Explosions, Useful Formulæ, etc. With 36 diagrams and 4 plates. Second edition, revised and enlarged. 12mo, cloth............... 1 50

—— Marine Boilers: A Treatise on the Causes and Prevention of their Priming, with Remarks on their General Management. Illustrated. 12mo, cloth... 2 00

REINHARDT (CHAS. W.). Lettering for Draftsmen, Engineers and Students. A Practical System of Free-hand Lettering for Working Drawings. Thoroughly revised and largely rewritten. Thirteenth thousand. Oblong, boards.................. 1 00

RICE (J. M.)., and JOHNSON, W. W. On a New Method of obtaining the Differential of Functions, with especial reference to the Newtonian Conception of Rates or Velocities. 12mo, paper.... 50

RINGWALT (J. L.). Development of Transportation Systems in the United States, Comprising a Comprehensive Description of the leading features of advancement from the colonial era to the present time. With illustrations. Quarto, half morocco.. 7 50

RIPPER (WILLIAM). A Course of Instruction in Machine Drawing and Design for Technical Schools and Engineer Students. With 52 plates and numerous explanatory engravings. Folio, cloth.......... 6 00

ROEBLING (J. A.). Long and Short Span Railway Bridges. Illustrated with large copperplate engravings of plans and views. Imperial folio, cloth..25 00

SCIENTIFIC PUBLICATIONS. 31

ROGERS (Prof. H. D.). The] Geology of Pennsylvania. A Government Survey, with a General View of the Geology of the United States, essays on the Coal Formation and its Fossils, and a description of the Coal Fields of North America and Great Britain. Illustrated with plates and engravings in the text. 3 vols., 4to, cloth, with portfolio of maps. 15 00

ROSE (JOSHUA, M. E.). The Pattern-Makers' Assistant. Embracing Lathe Work, Branch Work, Core Work, Sweep Work, and Practical Gear Constructions, the Preparation and Use of Tools, together with a large collection of useful and valuable Tables. Tenth edition. Illustrated with 250 engravings. 8vo, cloth 2 50

—— Key to Engines and Engine-running. A Practical Treatise upon the Management of Steam Engines and Boilers for the Use of Those who Desire to Pass an Examination to Take Charge of an Engine or Boiler. With numerous illustrations, and Instructions Upon Engineers' Calculations, Indicators, Diagrams, Engine Adjustments, and other Valuable Information necessary for Engineers and Firemen. 12mo, cloth.. 2 50

SABINE (ROBERT). History and Progress of the Electric Telegraph. With descriptions of some of the apparatus. Second edition, with additions. 12mo, cloth.. 1 25

SAELTZER (ALEX.). Treatise on Acoustics in connection with Ventilation. 12mo, cloth......... ..... 1 00

SALOMONS (Sir DAVID, M. A.). Electric Light Installations. A Practical Hand-book. Eighth edition, revised and enlarged, with numerous illustrations. Vol. I. The management of Accumulators. 12mo, cloth.. 1 50
Vol. II., Apparatus, 296 illustrations. 12mo, cloth. 2 25
Vol. III., Applications. 12mo, cloth............ 1 50

SANFORD (P. GERALD). Nitro-Explosives. A Practical Treatise concerning the Properties, Manufacture and Analysis of Nitrated Substances, including the Fulminates, Smokeless Powders and Celluloid. 8vo, cloth, 270 pages............ .... 3 00

SAUNNIER (CLAUDIUS). Watchmaker's Handbook. A Workshop Companion for those engaged in Watchmaking and allied Mechanical Arts. Translated by J. Tripplin and E. Rigg. Second edition, revised with appendix. 12mo, cloth............ 3 50

SCHELLEN (Dr. H.). Magneto-Electric and Dynamo-Electric Machines: their Construction and Practical Application to Electric Lighting, and the Transmission of Power. Translated from the third German edition, by N. S. Keith and Percy Neymann, Ph. D. With very large additions and notes relating to American Machines, by N. S. Keith. Vol. I., with 353 illustrations. Second edition..................... 5 00

SCHUMANN (F.). A Manual of Heating and Ventilation in its Practical Application, for the use of Engineers and Architects. Embracing a series of Tables and Formulæ for dimensions of heating, flow and return pipes for steam and hot-water boilers, flues, etc. 12mo, illustrated, full roan. 1 50

SCRIBNER (J. M.). Engineers' and Mechanics' Companion. Comprising United States Weights and Measures. Mensuration of Superfices and Solids, Tables of Squares and Cubes, Square and Cube Roots, Circumference and Areas of Circles, the Mechanical Powers, Centres of Gravity, Gravitation of Bodies, Pendulums, Specific Gravity of Bodies, Strength, Weight, and Crush of Materials, Water-Wheels, Hydrostatics, Hydraulics, Statics, Centres of Percussion and Gyration, Friction Heat, Tables of the Weight of Metals, Scantling, etc., Steam and the Steam Engine. Twentieth edition revised. 16mo, full morocco.. 1 50

SEATON (A. E.). A Manual of Marine Engineering. Comprising the Designing, Construction and Working of Marine Machinery. With numerous tables and illustrations reduced from Working Drawings. Fourteenth edition. Revised throughout, with an additional chapter on Water Tube Boilers. 8vo., cloth... 6 00

—— and ROUNTHWAITE (H. M.). A Pocketbook of Marine Engineering Rules and Tables. For the use of Marine Engineers and Naval Architects, Designers, Draughtsmen, Superintendents, and all engaged in the design and construction of Marine Machinery, Naval and Mercantile. Fifth edition, revised and enlarged. Pocket size. Leather, with diagrams. 12mo. morocco Illustrated............ 3 00

SEXTON (A. HUMBOLDT). Fuel and Refractory
Materials. 8vo, cloth.................................. 2 00

SHIELDS (J. E.). Notes on Engineering Construction. Embracing Discussions of the Principles involved, and Descriptions of the Material employed in Tunnelling, Bridging, Canal and Road Building, etc. 12mo, cloth............................... 1 50

SHOCK (WM. H.). Steam Boilers: Their Design, Construction and Management. 4to, half morocco.15 00

SHREVE (S. H.). A Treatise on the Strength of Bridges and Roofs. Comprising the determination of Algebraic formulas for strains in Horizontal, Inclined or Rafter, Triangular, Bowstring, Lenticular, and other Trusses, from fixed and moving loads, with practical applications and examples, for the use of Students and Engineers. 87 woodcut illus. Fourth edition. 8vo, cloth 3 50

SHUNK (W. F.). The Field Engineer. A Handy Book of practice in the Survey, Location, and Truckwork of Railroads, containing a large collection of Rules and Tables, original and selected, applicable to both the Standard and Narrow Gauge, and prepared with special reference to the wants of the young Engineer. Fourteenth edition, revised and enlarged. 12mo, morocco, tucks................. 2 50

SIMMS (F. W.). A Treatise on the Principles and Practice of Levelling. Showing its application to purposes of Railway Engineering, and the Construction of Roads, etc. Revised and corrected, with the addition of Mr. Laws' Practical Examples for setting out Railway Curves. Illustrated. 8vo, cloth.. 2 50

IMMS (W. F.). Practical Tunnelling. Fourth edition, revised and greatly extended. With additional chapters illustrating recent practice by D. Kinnear Clark. With 36 plates and other illustrations. Imperial 8vo, cloth..................................12 00

SLATER (J. W.). Sewage Treatment, Purification, and Utilization. A Practical Manual for the Use of Corporations, Local Boards, Medical Officers of Health, Inspectors of Nuisances, Chemists, Manufacturers, Riparian Owners, Engineers, and Ratepayers. 12mo, cloth............................. 2 25

SMITH (ISAAC W., C. E.). The Theory of Deflections and of Latitudes and Departures. With special applications to Curvilinear Surveys, for Alignments of Railway Tracks. Illustrated. 16mo, morocco, tucks.. 3 00

SNELL (ALBION T.). Electric Motive Power: The Transmission and Distribution of Electric Power by Continuous and Alternate Currents. With a Section on the Applications of Electricity to Mining Work. Second edition. 8vo, cloth, illustrated.............. 4 00

SPEYERS (CLARENCE L.). Text-Book of Physical Chemistry. 8vo, cloth................ 2 25

STAHL (A. W.), and A. T. WOODS. Elementary Mechanism. A Text-Book for Students of Mechanical Engineering. Eleventh edition, enlarged. 12mo, cloth.. 2 00

STALEY (CADY), and PIERSON, GEO. S. The Separate System of Sewerage: its Theory and Construction. Third edition, revised. 8vo, cloth. With maps, plates and illustrations......................... 3 00

STEVENSON (DAVID, F.R.S.N.). The Principles and Practice of Canal and River Engineering. Revised by his sons David Alan Stevenson, B. Sc., F. R. S.E., and Charles Alexander Stevenson, B. Sc., F.R. S.E., Civil Engineer. Third edition, with 17 plates, 8vo, cloth......................................10 00

—— The Design and Construction of Harbors, A Treatise on Maritime Engineering. Third edition with 24 plates, 8vo, cloth......................10 00

STEWART (R. W.). A Text Book of Light, Adapted to the Requirements of the Intermediate Science and Preliminary Scientific Examinations of the University of London, and also for General Use, Numerous Diagrams and Examples. 12mo, cloth.. 1 00

STEWART (R. W.). A Text Book of Heat, Illustrated, 8vo, cloth.............................. 1 00

—— A Text Book of Magnetism and Electricity, 160 Illus. and Numerous Examples. 12mo, cloth....... 1 00

STILES (AMOS). Tables for Field Engineers. Designed for use in the field. Tables containing all the functions of a one degree curve, from which a corresponding one can be found for any required

degree. Also, Tables of Natural Sines and Tangents.
12mo, morocco, tucks............................... 2 00

STILLMAN (PAUL). Steam Engine Indicator and the Improved Manometer Steam and Vacuum Gauges; their Utility and Application. New edition. 12mo, flexible cloth.............................. 1 00

STONE (General ROY). New Roads and Road Laws in the United States. 200 pages, with numerous illustrations. 12mo, cloth........................ 1 00

STUART (C. B., U. S. N.). Lives and Works of Civil and Military Engineers of America. With 10 steel-plate engravings. 8vo, cloth........................ 5 00

—— The Naval Dry Docks of the United States. Illustrated with 24 fine Engravings on Steel. Fourth edition. 4to, cloth........................ 6 00

SWINTON (ALAN A. CAMPBELL). The Elementary Principle of Electric Lighting. Illustrated. 12mo, cloth 60

TEMPLETON (WM.). The Practical Mechanic's Work-shop Companion. Comprising a great variety of the most useful rules and formulæ in Mechanical Science, with numerous tables of practical data and calculated results facilitating mechanical operations. Revised and enlarged by W. S. Hutton. 12mo, morocco........................ 2 00

THOM (CHAS.), and WILLIS H. JONES. Telegraphic Connections: embracing Recent Methods in Quadruplex Telegraphy. Oblong, 8vo, cloth. 20 full page plates, some colored.................. 1 50

THOMPSON (EDWARD P., M. E.). How to Make Inventions; or Inventing as a Science and an Art. A Practical Guide for Inventors. Second, edition. 8vo, boards 1 00

—— Roentgen Rays and Phenomena of the Anode and Cathode Principles, Applications and Theories. For Students, Teachers, Physicians, Photographers, Electricians and others. Assisted by Louis M. Pignolet, N. D. C. Hodges, and Ludwig Gutmann, E. E. With a Chapter on Generalizations, Arguments, Theories, Kindred Radiations and Phenomena. By Professor Wm. Anthony. 8vo, cloth. 50 Diagrams, 40 Half tones..................... 1 50

TODD (JOHN), and W. B. WHALL. Practical Seamanship for Use in the Merchant Service: Including all ordinary subjects; also Steam Seamanship, Wreck Lifting, Avoiding Collision, Wire Splicing, Displacement, and everything necessary to be known by seamen of the present day. Second edition, with 247 illustrations and diagrams. 8vo, cloth.......... 8 00

TOOTHED GEARING. A Practical Hand-book for Offices and Workshops. By a Foreman Patternmaker. 184 Illustrations. 12mo, cloth............. 2 25

TRATMAN (E. E. RUSSELL). Railway Track and Track Work. With over two hundred illustrations. 8vo, cloth.. ⁻ 00

TREVERT (EDWARD). How to Build Dynamo Electric Machinery, embracing Theory Designing and Construction of Dynamos and Motors. With appendices on Field Magnet and Armature Winding, Management of Dynamos and Motors, and Useful Tables of Wire Gauges. Illustrated. 8vo, cloth ... 2 50

—— Electricity and its Recent Applications. A Practical Treatise for Students and Amateurs, with an Illustrated Dictionary of Electrical Terms and Phrases. Illustrated. 12mo, cloth.................. 2 00

TUCKER (Dr. J. H.). A Manual of Sugar Analysis, including the Applications in General of Analytical Methods to the Sugar Industry. With an Introduction on the Chemistry of Cane Sugar. Dextrose, Levulose, and Milk Sugar. 8vo, cloth. Illustrated. 3 50

TUMLIRZ (Dr. O.). Potential and its Application to the Explanation of Electric Phenomena, Popularly Treated. Translated from the German by D. Robertson. Illustrated. 12mo, cloth................. .... 1 25

TUNNER (P. A.). Treatise on Roll-Turning for the Manufacture of Iron. Translated and adapted by John B. Pearse, of the Pennsylvania Steel Works, with numerous engravings, wood-cuts. 8vo, cloth, with folio atlas of plates...................... .10 00

URQUHART (J. W.). Electric Light Fitting. Embodying Practical Notes on Installation Management. A Hand-book for Working Electrical Engineers— with numerous illustrations. 12mo, cloth.. .... 2 00

—— Electro-Plating. A Practical Hand-book on the Deposition of Copper, Silver, Nickel, Gold, Brass, Aluminum, Platinum, etc. Fourth edition. 12mo.. 2 00

SCIENTIFIC PUBLICATIONS. 37

URQUHART, (J. W.). Dynamo Construc-tion : a Practical Hand-book for the Use of Engineer Constructors and Electricians in Charge, embracing Frame Work Building, Field Magnet and Armature Winding and Grouping, Compounding, etc., with Examples of Leading English, American and Continental Dynamos and Motors, with numerous illustrations. 12mo, cloth............................ 3 00

—— Electric Ship Lighting. A Hand-book on the Practical Fitting and Running of Ship's Electrical Plant. For the Use of Ship Owners and Builders, Marine Electricians and Sea Going Engineers-in-Charge. Numerous illustrations. 12mo, cloth 3 00

UNIVERSAL TELEGRAPH CIPHER CODE. Arranged for General Correspondence. 12mo, cloth. 1 00

VAN NOSTRAND'S ENGINEERING MAGAZINE. Complete sets, 1869 to 1886 inclusive.
Complete sets, 35 vols., in cloth 60 00
Complete sets, 35 vols., in half morocco.......... 100 00

VAN WAGENEN (T. F.). Manual of Hydraulic Mining. For the Use of the Practical Miner. Revised and enlarged edition. 18mo, cloth......... 1 00

WALKER (SIDNEY F.). Electric Lighting for Marine Engineers, or How to Light a Ship by the Electric Light and How to Keep the Apparatus in Order. 103 illustrations. 8vo, cloth. Second edition. 2 00

WALLIS-TAYLER (A. J.). Modern Cycles. A Practical Hand-book on their Construction and Repair. With 300 illustrations. 8vo, cloth.......... 4 00

—— Motor Cars, or Power Carriages for Common Roads. 8vo, cloth, with numerous illustrations..... 1 80

—— Bearings and Lubrication. A Hand-book for every user of Machinery. 8vo, cloth, fully illustrated 1 50

—— Refrigeration and Cold Storage, and Ice-making. A practical treatise on the art and science of refrigeration. With 361 cuts and diagrams. 8vo. Cloth. Illustrated..net 4 50

—— Sugar Machinery. A Descriptive Treatise devoted to the Machinery and Apparatus used in the Manufacture of Cane and Beet Sugars. 12mo, cloth. Illustrated.. 2 00

WANKLYN (J. A.). A Practical Treatise on the Examination of Milk and its Derivatives, Cream, Butter, and Cheese. 12mo, cloth.................... 1 00
——— Water Analysis. A Practical Treatise on the Examination of Potable Water. Tenth Edition. 12mo, cloth... 2 00

WANSBROUGH (WM. D.). The A. B. C. of the Differential Calculus. 12mo, cloth............... 1 50

WARD (J. H.). Steam for the Million. A Popular Treatise on Steam, and its application to the Useful Arts, especially to Navigation. 8vo, cloth 1 00

WARING (GEO. E., Jr.). Sewerage and Land Drainage. Illustrated with wood-cuts in the text, and full-page and folding plates. Quarto. Cloth. Third edition...................................... 6 00
——— Modern Methods of Sewage Disposal for Towns, Public Institutions and Isolated Houses. Second edition, revised and enlarged. 260 pages. Illustrated. Cloth.. 2 00
——— How to Drain a House. Practical Information for Householders. New and enlarged edition. 12mo, cloth... 1 25

WATSON (E. P.). Small Engines and Boilers. A Manual of Concise and Specific Directions for the Construction of Small Steam Engines and Boilers of Modern Types from five Horse-power down to model sizes. 12mo, cloth. Illustrated with Numerous Diagrams and Half Tone Cuts. New York, 1899....... 1 25

WATT (ALEXANDER). The Electro-plating and Electro-refining of Metals: being a new edition of Alexander Watt's "Electro-deposition." Revised and largely rewritten by Arnold Philip, B.Sc. With numerous figures and engravings. 8vo. Cloth. Illustrated ....net 4 50
——— Electro-Metallurgy Practically Treated. Tenth edition, considerably enlarged. 12mo, cloth. 1 00
——— The Art of Soap-Making. A Practical Hand-book of the Manufacture of Hard and Soft Soaps, Toilet Soaps, &c. Including many New Processes, and a Chapter on the Recovery of Glycerine from Waste Leys. With illustrations. Fourth edition, revised and enlarged. 8vo.................................... 3 00

WATT (ALEXANDER). The Art of Leather Manufacture. Being a Practical Hand-book, in which the Operations of Tanning, Currying, and Leather Dressing are Fully Described, and the Principles of Tanning Explained, and many Recent Processes Introduced. With numerous illustrations. Second edition. 8vo, cloth 4 00

WEALE (JOHN). A Dictionary of Terms Used in Architecture, Building, Engineering, Mining, Metallurgy, Archaeology, the Fine Arts etc., with explanatory observations connected with applied Science and Art. Fifth edition, revised and corrected. 12mo, cloth 2 50

—— Weale's Rudimentary Scientific Series (Catalogue sent on application).

WEBB (HERBERT LAWS). A Practical Guide to the Testing of Insulated Wires and Cables. Illustrated. 12mo, cloth 1 00

—— The Telephone Hand-book. 128 illustrations. 146 pages. 16mo, cloth 1 00

WEEKES (R. W.). The Design of Alternate Current Transformers. Illustrated. 12mo, cloth 1 00

WEISBACH (JULIUS). A Manual of Theoretical Mechanics. Ninth American edition. Translated from the fourth augmented and improved German edition, with an Introduction to the Calculus by Eckley B. Coxe, A. M., Mining Engineer. 1,100 pages, and 902 wood-cut illustrations. 8vo, cloth 6 00
Sheep 7 50

WESTON (EDMUND B.). Tables Showing Loss of Head Due to Friction of Water in Pipes. Second edition. 12mo, leather 1 50

WEYMOUTH (F. MARTEN). Drum Armatures and Commutators. (Theory and Practice.) A complete Treatise on the Theory and Construction of Drum Winding, and of commutators for closed-coil armatures, together with a full resume of some of the principal points involved in their design, and an exposition of armature re-actions and sparking. 8vo, cloth 3 00

WHITE (W. H., K.C.B.). A Manual of Naval Architecture, for use of Officers of the Royal Navy, Officers of the Mercantile Marine, Yachtsmen, Shipowners and Shipbuilders. Containing many figures, diagrams and tables. Thick 8vo, cloth, illus 9 00

WHEELER (Prof. J. B.). Art of War. A Course of Instruction in the Elements of the Art and Science of War, for the Use of the Cadets of the United States Military Academy, West Point, N. Y. 12mo, cloth.... .. .. 1 75
—— Field Fortifications. The Elements of Field Fortifications, for the Use of the Cadets of the United States Military Academy, West Point, N. Y. 12mo. 1 75
WHIPPLE (S., C. E.). An Elementary and Practical Treatise on Bridge Building. 8vo, cloth............ 3 00
WILKINSON (H. D.). Submarine Cable-Laying, Repairing and Testing. 8vo, cloth......... ...... 4 00
WILLIAMSON (R. S.). On the Use of the Barometer on Surveys and Reconnoissances. Part I. Meteorology in its Connection with Hypsometry. Part II. Barometric Hypsometry. With illustrative tables and engravings, 4to, cloth...... . ... 15 00
WILLIAMSON (E. S.). Practical Tables in Meteorology and Hpsometry, in connection with the use of the Barometer. 4to, cloth............................ 2 50
WILSON (GEO.). Inorganic Chemistry, with New Notation. Revised and enlarged by H. G. Madan. New edition. 12mo, cloth 2 00
WOODBURY (D. V.). Treatise on the Various Elements of Stability in the Well-Proportioned Arch. 8vo, half morocco..... .... 4 00
WRIGHT (T. W.). A Treatise on the Adjustment of Observations. With applications to Geodetic Work, and other Measures of Precision. 8vo, cloth........ 4 00
—— Elements of Mechanics; including Kinematics, Kinetics and Statics. With application. 8vo, cloth.. 2 50
WYLIE (CLAUDE). Iron and Steel Founding. Illustrated with 39 diagrams. Second edition, revised and enlarged. 8vo, cloth... .... 2 00
YOUNG (J. ELTON). Electrical Testing for Telegraph Engineers, with Appendices consisting of Tables. 8vo, cloth. Illustrated........ .... 4 00
YOUNG SEAMAN'S MANUAL. Compiled from Various Authorities, and Illustrated with Numerous Original and Select Designs, for the Use of the United States Training Ships and the Marine Schools. 8vo, half roan......................... 3 00
ZIPSER (JULIUS). Textile Raw Materials, and their Conversion into Yarns. Translated from the German by Chas. Salter. 8vo, cloth. Illustrated...... 5 00

www.ingramcontent.com/pod-product-compliance
Lightning Source LLC
Chambersburg PA
CBHW031830230426
43669CB00009B/1290